ESTIMATION THEORY
AND APPLICATIONS

ESTIMATION THEORY
AND APPLICATIONS

Nasser E. Nahi

Associate Professor of Electrical Engineering
University of Southern California
Los Angeles, California

JOHN WILEY & SONS, INC.   NEW YORK   LONDON   SYDNEY   TORONTO

To Mitra

PREFACE

Estimation, the process of extracting information concerning a parameter or signal function from noise-corrupted observations (experimental data), is very old. The concepts of least square estimation and curve fitting were introduced in the early 1800's by Legendre and Gauss mainly for the purpose of reducing physical and astronomical data. However, most major contributions to the field have been made in recent years. Because of the accessibility of computers and the consequent acceptability of otherwise long and tedious estimation algorithms, old problems have been reformulated in a setting appropriate for obtaining efficient recursive numerical solutions. As a result, recursive estimators now constitute a major portion of all estimators. Furthermore, it has become possible to implement methods of estimation which are, for certain problems, theoretically optimal. Such methods have become much more important because now it is computationally feasible to carry out the required algorithms on computers.

The derivations and applications of modern estimators and estimation algorithms are buried, so to speak, in the technical literature on communication theory, statistics, control theory, and others. Thus, it has been difficult for practicing engineers and scientists to assemble a comprehensive summary of useful results.

This book is intended as an introduction to the theory of estimation. Its purpose is to unify many of the results from the literature, establish consistent notation, clarify the assumptions and limitations of various techniques, and convey an understanding of the interrelationships of these techniques. After a brief review of pertinent mathematical fundamentals, followed by a discussion of the philosophy and basic methodology underlying most estimation methods, the book proceeds to the derivation of many types of classical and modern estimators. Emphasis is placed on recursive estimation techniques, wherever applicable, because of their great practical importance. Numerous examples are provided to help clarify theory and application. The book should be easily understood by first-year graduate students in engineering, mathematics, physics, or related areas. Practicing engineers will find the book useful, both as a text and as a reference, since most of the results are given in a form that permits direct implementation on a computer.

Chapters 1, 2, and 3 provide most of the necessary mathematical and conceptual background. Chapters 4 through 11 are self-contained treatments of principal estimation methods, and need not be covered in the order presented. The author believes, however, that the arrangement of chapters is a natural one, and it is recommended that this arrangement be used whenever the course objective is to convey a broad familiarity with the theory and the interrelationships among the various estimation techniques. A brief outline of the content of the chapters follows.

Chapter 1 provides a brief survey of various mathematical subjects employed in the book. Part or all of this chapter may be by-passed in a graduate course, or simply assigned as preparatory reading, depending on the level of students' background. Although Chapter 1 is not intended to be used as a sole source for gaining the required mathematical knowledge, the topics covered are self-contained, with necessary derivations, proofs, and examples, so that the only prerequisite is a certain degree of mathematical maturity. The concepts of estimation and the various types of estimators in use are explored in Chapter 2. Appropriate mathematical models for stochastic (Markov) processes and their significance are discussed in Chapter 3. Linear estimators and their recursive (Kalman) formulations are covered in Chapters 4, 5, and 6. Chapters 7 and 8 are mainly devoted to estimators applicable to nonlinear process models. The connection between "system identification" and estimation theory is established in Chapter 9, where appropriate estimators are derived. The connection with detection theory is brought out in Chapter 10, where the concept of the "matched filter" is introduced. Finally, Chapter 11 is devoted to analysis as well as further optimization and evaluation of estimation processes by means of the important Cramér-Rao lower bound for the variance of an estimate.

Pertinent references will be found at the end of each chapter. No attempt has been made to provide a complete bibliography.

It was decided, in the presentation, to concentrate on estimation procedures that are of a general nature and have had broad application. Unfortunately, because of space limitations, many elegant estimators applicable only to very restricted situations had to be omitted or only briefly considered.

The author is indebted to Professor Z. A. Kaprielian for his encouragement during the preparation of the manuscript. The proofreading was the work of Dr. D. Wallis, who also contributed innumerable constructive criticisms for improvement of the presentation.

NASSER E. NAHI

Los Angeles, California
January 1969

CONTENTS

chapter 1/A MATHEMATICAL SURVEY

1.1 Introduction

Historically the subject of estimation theory originated within the broad area of statistics and gradually found its way through many disciplines of science and engineering. Today the extent to which it has influenced a variety of subjects can be felt by enumerating the many seemingly unrelated areas of application such as satellite orbit determination, mathematical modeling of a human operator, optimal and adaptive control systems, determination of radar range, and many others. Essentially, estimation theory deals with statistical phenomena and dynamical behavior of systems of various types where the complexity of operations will, in general, necessitate their characterization by means of multidimensional variables. Consequently, the related areas in mathematics play a great role in achieving the desired results. A brief survey of these topics is the subject of this chapter. The material is grouped into five categories: matrix algebra and quadratic forms, vector differential and difference equations, probability theory, transform theory, and calculus of variations. No attempt is made to cover the various subjects with any degree of thoroughness. Instead, the emphasis is only on those subjects and results which are necessary for the development in subsequent chapters.

The purpose of this chapter, mainly, is to convey to the reader the extent to which he needs to be familiar with the mathematical notations and

1

techniques used in the remaining chapters. Thus, this chapter can serve as a review of these subjects if the reader has already been introduced to them. However, as the topics covered are self-contained, a certain degree of mathematical maturity is the only prerequisite for readers who have not previously studied these topics.

MATRIX ALGEBRA
AND QUADRATIC FORMS

1.2 Matrices

Let us start by introducing the concept of a matrix. An array of numbers written in the form

$$
A = \begin{bmatrix}
a_{11} & a_{12} & \cdots & a_{1n} \\
a_{21} & a_{22} & \cdots & a_{2n} \\
\cdot & & & \\
\cdot & & & \\
\cdot & & & \\
a_{m1} & \cdots & & a_{mn}
\end{bmatrix} \tag{1.1}
$$

is called a matrix of order $m \times n$. It is called a square matrix if $m = n$. The terms a_{ij} are referred to as elements of the matrix and in general can be functions (e.g., of time). When $n = 1$, the matrix reduces to a "column vector," and when $m = 1$, it is called a "row vector." When $n = m = 1$, the matrix becomes a scalar. The determinant associated with a square matrix A will be denoted by $|A|$. A square matrix where $a_{ii} = 1, i = 1, \ldots, n$ and $a_{ij} = 0$, $i \neq j$ is denoted by I and is called the identity matrix. If only $a_{ij} = 0$, $i \neq j$ the square matrix is called a diagonal matrix.

1.3 Matrix Operations

The following are the definitions of various matrix operations:

Equality. Two matrices A and B are equal if they have the same number of rows and columns and if the corresponding elements are equal, i.e.,

$$
a_{ij} = b_{ij}
$$

Addition. The sum of two matrices, A and B having the same dimensions $m \times n$ is defined by an $m \times n$ matrix C where

$$
c_{ij} = a_{ij} + b_{ij} \tag{1.2}
$$

Scalar Multiplication. If α is a scalar, the product αA is a matrix with elements αa_{ij}.

Matrix Multiplication. If A is an $n \times r$ matrix and B an $r \times m$ matrix, the product AB is given by an $n \times m$ matrix C where

$$c_{ij} = \sum_{k=1}^{r} a_{ik} b_{kj} \tag{1.3}$$

It can easily be shown that the operation of matrix multiplication is associative, i.e.,

$$(AB)C = A(BC)$$

where the parentheses indicate the multiplication to be carried out first. Matrix multiplication is not in general commutative, even in the case of square matrices. However, multiplication of diagonal matrices (which must be of the same dimension) is always commutative. In any matrix product AB, the matrix A is said to "premultiply" B, and the matrix B is said to "postmultiply" A.

Transposition. The transpose of an $n \times m$ matrix A is denoted by an $m \times n$ matrix A' with elements a'_{ij} where

$$a'_{ij} = a_{ji}$$

The transpose of the product of matrices A and B is given by

$$(AB)' = B'A'$$

By induction, it can be proved for any finite number of matrices $A, B, C, \ldots$ for which the product $ABC \cdots$ exists that

$$(ABC\cdots)' = \cdots C'B'A'$$

Linear Transformation. Multiplication of an $n \times 1$ vector x by an $m \times n$ matrix T is called a linear transformation or mapping. The resultant product

$$y = Tx$$

is an m-vector y. The transformation T is called a linear mapping, since if z is also an n-vector and α and β are scalars we have

$$T(\alpha x + \beta z) = \alpha Tx + \beta Tz$$

Inversion. The inverse of a square matrix A (if it exists) is denoted by A^{-1} and has the property that

$$A^{-1}A = AA^{-1} = I$$

where I is the identity matrix of the same dimension as A. The inverse of a matrix exists if and only if its determinant is nonzero. If we denote the elements of A^{-1} by α_{ij}, it can be shown that

$$\alpha_{ij} = \frac{(-1)^{i+j}|A|_{j,i}}{|A|} \qquad (1.4)$$

where $|A|_{j,i}$ is the determinant of the matrix A when the jth row and ith column have been dropped out. $|A|_{j,i}$ is called the cofactor of the term a_{ji}. The following is an example.

If

$$A = \begin{bmatrix} 1 & 2 \\ 3 & 4 \end{bmatrix}$$

then

$$A^{-1} = \begin{bmatrix} -2 & +1 \\ 3/2 & -1/2 \end{bmatrix}$$

A matrix which does not have an inverse (i.e., its determinant is zero) is called a singular matrix. It can easily be shown that if A and B are non-singular matrices then

$$(AB)^{-1} = B^{-1}A^{-1}$$

which is a relation of the same form as that given for the transpose. Furthermore, for any finite number of nonsingular matrices $A, B, C, \ldots$ it can be proved by induction that

$$(ABC\cdots)^{-1} = \cdots C^{-1}B^{-1}A^{-1}$$

1.4 Eigenvalues and Eigenvectors

The eigenvalues of any $n \times n$ matrix A are defined as the roots of the polynomial equation in λ given below

$$|A - \lambda I| = 0 \qquad (1.5)$$

where λ is a scalar and I is the identity matrix of the same dimension as A.

This polynomial equation is referred to as the characteristic equation of the matrix, and is of order n. Hence any matrix of order n has n eigenvalues $\lambda_i, i = 1, \ldots, n$. The eigenvalues can in general be real or complex quantities. If $\lambda_i \neq \lambda_j, i \neq j$, the eigenvalues are said to be distinct.

The eigenvectors of a matrix A are defined as the *nonzero* vector solutions, for each λ_i, of the equations

$$Ax = \lambda_i x, \qquad i = 1, \ldots, n \qquad (1.6)$$

An eigenvector x_i corresponding to an eigenvalue λ_i is defined only to within a constant multiplier because if x^i satisfies the equation $Ax^i = \lambda_i x^i$, then $A(\alpha x^i) = \lambda_i(\alpha x^i)$. Thus, we are free to choose α to give the eigenvector any magnitude we wish.

▶ **example**
The characteristic equation of the matrix

$$A = \begin{bmatrix} 2 & 1 \\ 3 & 4 \end{bmatrix}$$

is given by

$$A - \lambda I = \lambda^2 - 6\lambda + 5 = 0$$

yielding the eigenvalues $\lambda_1 = 1$, $\lambda_2 = 5$.

The eigenvectors are given by the solutions of the simultaneous equations given below.

$$2x^1 + x^2 = \lambda_i x^1$$
$$3x^1 + 4x^2 = \lambda_i x^2$$
$$i = 1, 2$$

resulting in

$$x^1 = \begin{bmatrix} 1 \\ -1 \end{bmatrix} \quad \text{and} \quad x^2 = \begin{bmatrix} 1 \\ 3 \end{bmatrix}$$

Because we are free to choose constant scalar multipliers α, we could also have obtained

▶
$$x^1 = \begin{bmatrix} -14 \\ 14 \end{bmatrix} \quad x^2 = \begin{bmatrix} 2 \\ 6 \end{bmatrix}$$

1.5 Matrix Rank and Linear Dependence

The columns (or rows) of a matrix are called linearly dependent if any column (or row) can be written as a linear combination of the remaining columns (rows). For example, in the matrix

$$A = \begin{bmatrix} 1 & 1 & 0 \\ 3 & 1 & 1 \\ 5 & 1 & 2 \end{bmatrix}$$

where the columns are

$$c^1 = \begin{bmatrix} 1 \\ 3 \\ 5 \end{bmatrix}$$

$$c^2 = \begin{bmatrix} 1 \\ 1 \\ 1 \end{bmatrix}$$

$$c^3 = \begin{bmatrix} 0 \\ 1 \\ 2 \end{bmatrix}$$

it is clearly seen that

$$c^1 = c^2 + 2c^3$$

Or, if we denote the rows by r^1, r^2, and r^3, then we have

$$r^3 = -r^1 + 2r^2$$

It is evident that when the rows (or columns) of a square matrix are linearly dependent, the determinant of the matrix is zero, and thus the matrix is singular. In fact, a necessary and sufficient condition for a matrix to be nonsingular is that all its columns (or, equivalently, rows) be linearly independent. Furthermore, when the rows of a square matrix are linearly dependent, so are the columns, and vice versa.

The rank of a matrix is the largest number of columns (rows) which are linearly independent. For example, an $n \times n$ nonsingular matrix has rank n, and the following matrix has rank 2.

$$A = \begin{bmatrix} 1 & 0 & 0 \\ 0 & 0 & 0 \\ 0 & 0 & 3 \end{bmatrix}$$

Furthermore, it will be shown later that the rank of a matrix is equal to the number of its nonzero eigenvalues.

1.6 Inner and Outer Vector Products

Let x and y be two n-dimensional column vectors. The scalar quantity

$$x'y = \sum_{i=1}^{n} x_i y_i \tag{1.7}$$

is called the inner (or dot) product of the two vectors. The inner product of a vector with itself is the square of the (so-called "Euclidean") length of the vector

$$x'x = \sum_{i=1}^{n} x_i^2 \tag{1.8}$$

This is the main reason for the importance of the inner product. The notation $\|x\|$ will be used for denoting $[x'x]^{1/2}$, the Euclidean length or "norm" of the vector x.

A matrix T having the property that

$$\|Tx\| = \|x\| \qquad \text{for all } x \tag{1.9}$$

(i.e., the length of a vector is invariant through the linear mapping T) is called an orthogonal matrix. Now, for the matrix T above, we have

$$\|Tx\|^2 = x'T'Tx = x'x \tag{1.10}$$

Consequently a property of any orthogonal matrix T is that

$$T'T = I$$
$$T' = T^{-1} \tag{1.11}$$

An orthogonal matrix is never singular.

Clearly, since the inner product $x'y$ is a scalar quantity we have

$$x'y = y'x$$

If the inner product of two vectors is zero they are said to be orthogonal. This usage of the term "orthogonal" differs from that applied to transformations.

The outer product of two column n-vectors x and y is defined by the following $n \times n$ matrix

$$xy' = \begin{bmatrix} x_1y_1 & x_1y_2 & \cdots & x_1y_n \\ x_2y_1 & x_2y_2 & \cdots & \\ \cdot & & & \\ \cdot & & & \\ \cdot & & & \\ x_ny_1 & \cdots & & x_ny_n \end{bmatrix} \tag{1.12}$$

It is easy to establish that this matrix always has rank 1.

1.7 Symmetric Matrices

The matrix A is symmetric if it is square and if

$$a_{ij} = a_{ji}$$

Symmetric matrices play an important role in estimation theory and, for that matter, in various branches of science and engineering. They possess many interesting properties, among which are the following:

The eigenvalues of a symmetric matrix with real coefficients are real. To prove this assertion, let us assume the contrary, that is, there exists a complex root λ of the characteristic equation of matrix A

$$|A - \lambda I| = 0 \qquad (1.13)$$

Since the characteristic equation is a polynomial in λ with real coefficients, the complex conjugate of λ, denoted by $\bar{\lambda}$, will also have to be another of its solutions. If x is the eigenvector associated with λ, and $\bar{x}$ is the eigenvector associated with $\bar{\lambda}$, we have

$$Ax = \lambda x$$
$$A\bar{x} = \bar{\lambda}\bar{x}$$

Premultiplying these equations by $\bar{x}'$ and x' respectively yields

$$\bar{x}'Ax = \lambda \bar{x}'x$$
$$x'A\bar{x} = \bar{\lambda}x'\bar{x}$$

Since A is symmetric

$$(x'A\bar{x})' = \bar{x}'A'x = \bar{x}'Ax$$

it follows that

$$\lambda = \bar{\lambda}$$

which is a contradiction.

Another interesting property of a symmetric matrix is that the eigenvectors associated with distinct eigenvalues are orthogonal. Let λ_1 and λ_2 be two distinct eigenvalues of the matrix A. Let the vectors x^1 and x^2 denote the corresponding eigenvectors. By definition,

$$Ax^1 = \lambda_1 x^1$$
$$Ax^2 = \lambda_2 x^2$$

Hence

$$x^{2\prime}Ax^1 = \lambda_1 x^{2\prime}x^1$$
$$x^{1\prime}Ax^2 = \lambda_2 x^{1\prime}x^2$$

Since $x^{1\prime}Ax^2 = x^{2\prime}Ax^1$ and $x^{2\prime}x^1 = x^{1\prime}x^2$, subtracting the second equation from the first yields

$$0 = (\lambda_1 - \lambda_2)x^{2\prime}x^1$$

Now if $\lambda_1 \neq \lambda_2$ (i.e., eigenvalues are distinct) we then have

$$x^{2\prime}x^1 = 0$$

which establishes that x^1 and x^2 are orthogonal.

1.8 Matrix Diagonalization

Given an arbitrary $n \times n$ matrix A with distinct eigenvalues, there exists a nonsingular $n \times n$ matrix P such that the matrix

$$P^{-1}AP$$

is diagonal. Defining this diagonal matrix by

$$\Lambda \triangleq \text{diag}\,[\lambda_1, \ldots, \lambda_n]$$

we have then

$$AP = P\,\text{diag}\,[\lambda_1, \ldots, \lambda_n] = P\Lambda \qquad (1.14)$$

or in expanded form

$$\sum_{j=1}^{n} a_{ij} p_{jk} = p_{ik}\lambda_k \qquad \begin{matrix} i = 1, \ldots, n \\ k = 1, \ldots, n \end{matrix}$$

Defining the kth column of matrix P by p^k, i.e.,

$$p^k = \begin{bmatrix} p_{1k} \\ p_{2k} \\ \cdot \\ \cdot \\ \cdot \\ p_{nk} \end{bmatrix}$$

we have

$$Ap^k = \lambda_k p^k, \qquad k = 1, \ldots, n \qquad (1.15)$$

Consequently the determination of matrices P and diag $[\lambda_1, \ldots, \lambda_n]$ reduces to finding vectors p^k such that the transformation through A is the same as multiplication by a constant λ_k. Clearly, eigenvectors of A satisfy this condition.

▶ example
The eigenvalues and eigenvectors of matrix A given by

$$A = \begin{bmatrix} 2 & 1 \\ 3 & 4 \end{bmatrix}$$

were obtained in Section 1.4 as

$$\lambda_1 = 1 \qquad \lambda_2 = 5$$

$$x^1 = \begin{bmatrix} 1 \\ -1 \end{bmatrix} \qquad x^2 = \begin{bmatrix} 1 \\ 3 \end{bmatrix}$$

Consequently, the diagonalization matrix P is given by

$$P = \begin{bmatrix} 1 & 1 \\ -1 & 3 \end{bmatrix}$$

The following is easily shown by carrying out the multiplications.

$$P^{-1}AP = \begin{bmatrix} 1 & 0 \\ 0 & 5 \end{bmatrix} = \Lambda$$

Note that if the columns of P had been reversed, the order of the eigenvalues would have been altered, so that

▶
$$P^{-1}AP = \begin{bmatrix} 5 & 0 \\ 0 & 1 \end{bmatrix}$$

If matrix A is symmetric with all of its eigenvalues distinct* and if we construct P by choosing its columns to be equal to the eigenvectors normalized to have unit length, then this diagonalizing matrix P will be an orthogonal matrix. To verify this, recall that we established in Section 1.7 that the eigenvectors of a symmetric matrix corresponding to distinct eigenvalues are orthogonal. It immediately follows that $P'P$ is a diagonal matrix. Furthermore, each element on the main diagonal is the square of the length of an eigenvector, each of which was chosen here as unity. Thus, $P'P = I$, which establishes that P is an orthogonal matrix.

▶ **example**
Let the matrix A be given by

$$A = \begin{bmatrix} -3 & 1 \\ 1 & -3 \end{bmatrix}$$

The eigenvalues are obtained as

$$\lambda_1 = -2, \qquad \lambda_2 = -4$$

The eigenvectors are given by

$$x^1 = \alpha \begin{bmatrix} 1 \\ 1 \end{bmatrix}$$

$$x^2 = \beta \begin{bmatrix} 1 \\ -1 \end{bmatrix}$$

* The assumption of distinct eigenvalues is not necessary. However, one then has to establish that to any eigenvalue of order r there correspond r orthogonal eigenvectors.

for any arbitrary value of α and β. Choosing α and β such that

$$x^{1'}x^1 = x^{2'}x^2 = 1$$

yields

$$x^1 = \begin{bmatrix} 1/\sqrt{2} \\ 1/\sqrt{2} \end{bmatrix}$$

$$x^2 = \begin{bmatrix} 1/\sqrt{2} \\ -1/\sqrt{2} \end{bmatrix}$$

Consequently, the diagonalization matrix becomes

$$P = \begin{bmatrix} \dfrac{1}{\sqrt{2}} & \dfrac{1}{\sqrt{2}} \\ \dfrac{1}{\sqrt{2}} & -\dfrac{1}{\sqrt{2}} \end{bmatrix}$$

That P diagonalizes A can be easily verified by performing the multiplication yielding

$$P^{-1}AP = \begin{bmatrix} -2 & 0 \\ 0 & -4 \end{bmatrix}$$

Furthermore, we have

$$P^{-1} = \begin{bmatrix} \dfrac{1}{\sqrt{2}} & \dfrac{1}{\sqrt{2}} \\ \dfrac{1}{\sqrt{2}} & -\dfrac{1}{\sqrt{2}} \end{bmatrix} = P'$$

▶ which verifies that $P^{-1} = P'$.

1.9 Quadratic Forms

A quadratic form is a scalar function of n variables $x_1, \ldots, x_n$ in the form

$$Q_n(x) = \sum_{j=1}^{n} \sum_{i=1}^{n} a_{ij} x_i x_j \tag{1.16}$$

A quadratic form is called positive definite if

$$Q_n(x) > 0 \quad \text{for all } x, \quad x \neq 0 \tag{1.17}$$

It is called positive semi-definite if

$$Q_n(x) \geq 0 \qquad \text{for all } x, \quad x \neq 0 \qquad (1.18)$$

A quadratic form is negative definite (semi-definite) if its negative is positive definite (semi-definite).

A convenient way of expressing quadratic forms is to use matrix notation. If A is an $n \times n$ matrix with elements a_{ij}, we can see that

$$Q_n(x) = x'Ax \qquad (1.19)$$

is a quadratic form.

A matrix A is called positive (or negative) definite (or semi-definite) if the quadratic form associated with it has the same property.

To any quadratic form $Q_n(x)$ there corresponds an infinite number of matrices A where (1.19) is satisfied. Among them there is always a symmetric matrix. If matrix A in (1.19) is not symmetric, it can be substituted by $\frac{1}{2}(A + A')$ without changing the quadratic form. Since symmetric matrices have useful properties, it is always helpful to associate such a matrix with any quadratic form.

The following property of a positive definite or semi-definite symmetric matrix A is of considerable interest:

A real symmetric matrix A according to Section 1.8 can be written

$$A = P'\Lambda P \qquad (1.20)$$

where $\Lambda = \text{diag } [\lambda_1, \ldots, \lambda_n]$ and λ_i, $i = 1, \ldots, n$ are the eigenvalues of A. Substituting for A in (1.19) yields

$$Q_n(x) = x'P'\Lambda Px \qquad (1.21)$$

Introducing the vector y as

$$y = Px \qquad (1.22)$$

we get

$$Q_n(x) = y'\Lambda y = \sum_{i=1}^{n} \lambda_i y_i^2 \qquad (1.23)$$

Since P is nonsingular, to every x there corresponds a vector y and vice versa. Equation (1.23) reveals that if $\lambda_i > 0$, $i = 1, \ldots, n$ then for every y (and hence every x) the quadratic form is positive. If λ_i's are positive or zero, there exist some vectors y for which $Q_n(x) > 0$ and some for which $Q_n(x) = 0$. If some λ_i's are positive and some negative, we can then easily choose vectors x (or correspondingly y) so that $Q_n(x) > 0$ or $Q_n(x) < 0$. This is summarized as follows: The symmetric matrix A is positive definite if all its eigenvalues are positive. It is positive semi-definite if all its eigenvalues are non-negative. Similar results can be stated for negative definite or semi-definite matrices. Clearly a positive (or negative) definite matrix is always nonsingular, since if A is positive definite

$$A^{-1} = [P'\Lambda P]^{-1} = P'\Lambda^{-1}P \tag{1.24}$$

and

$$\Lambda^{-1} = \text{diag } [1/\lambda_1, 1/\lambda_2, \ldots, 1/\lambda_n] \tag{1.25}$$

and the diagonal matrix Λ^{-1} exists since none of λ_i's are zero.

Another simple test for positive definiteness of a symmetric matrix is given by Sylvester's theorem, which is stated here without proof.

Sylvester's Theorem. A symmetric matrix A is positive definite (semi-definite) if and only if the determinants of all of its leading principal minors are positive (non-negative).

▶ **example**

The matrix

$$A = \begin{bmatrix} a_{11} & a_{12} \\ a_{12} & a_{22} \end{bmatrix}$$

is positive definite if

$$a_{11} > 0$$

$$a_{11}a_{22} - a_{22}{}^2 > 0$$

or equivalently if the roots of the characteristic equation

$$(a_{11} - \lambda)(a_{22} - \lambda) - a_{12}{}^2 = 0$$

▶ are positive.

The following results dealing with positive definite matrices are of considerable importance:

If A is any $n \times n$ nonsingular matrix the product $A'A$ is positive definite since its associated quadratic form can be written as

$$x'A'Ax = \|Ax\|^2 \tag{1.26}$$

which is positive for all $x \neq 0$. Notice that if A were singular, then there would exist a nonzero vector x where $Ax = 0$.

If A is positive definite and B is nonsingular, the matrix product

$$B'AB$$

is positive definite. Furthermore, if the columns of B are only linearly independent, the matrix product remains positive definite, even if B is not a square matrix.

1.10   Matrix Functions

The elements of a matrix in general can be functions of, say, time. As such, the matrix properties defined in the preceding section apply to the matrix evaluated at any given value of the variable. For example, when we state that the matrix

$$A(t) = \begin{bmatrix} 2 + \cos t & -\sin t \\ \sin t & 2 + \cos t \end{bmatrix} \tag{1.27}$$

is positive definite, it implies that by substituting all possible values for t the corresponding matrices (here infinite in number) are all positive definite.

The operations of derivatives and integrals applied to matrix or vector functions are defined as follows: The derivative of the matrix A with elements $a_{ij}(t)$ with respect to t is a matrix with elements $\dfrac{d}{dt} a_{ij}(t)$. The integral of a matrix with elements $a_{ij}(t)$ is a matrix with elements $\int a_{ij}(t)\, dt$. For example, the derivative of A given by (1.27) is

$$\frac{d}{dt} A(t) = \begin{bmatrix} -\sin t & -\cos t \\ \cos t & -\sin t \end{bmatrix} \tag{1.28}$$

1.11   Scalar Functions of a Square Matrix

In some applications it is helpful to have a scalar function defined by the elements of a square matrix. For example, the function

$$f(A) = \sum_{j=1}^{n} \sum_{i=1}^{n} |a_{ij}| \tag{1.29}$$

(where $|\cdot|$ denotes the absolute value) can be used to find an upper bound on the change in length of a vector introduced by a linear transformation A. Another important function of a square matrix is its *trace*, defined by

$$\operatorname{tr} A = a_{11} + a_{22} + \cdots + a_{nn} \tag{1.30}$$

Let the characteristic polynomial of A be given by

$$|\lambda I - A| = \lambda^n - \alpha_1 \lambda^{n-1} + \cdots + \alpha_n \tag{1.31}$$

On the other hand, if we expand the determinant $|A - \lambda I|$ we get

$$\alpha_1 = a_{11} + a_{22} + \cdots + a_{nn}$$

Hence*

$$\text{tr } A = \sum_{i=1}^{n} \lambda_i = \sum_{i=1}^{n} a_{ii} \tag{1.32}$$

Furthermore, it immediately follows for any two square matrices A and B that

$$\text{tr } (A + B) = \text{tr } A + \text{tr } B \tag{1.33}$$

and

$$\text{tr } (AB) = \text{tr } (BA) \tag{1.34}$$

It is interesting that these relationships hold even when there is no simple relation between the eigenvalues of $A + B$, A, B, AB, and BA.

1.12 Gradient Vector

The gradient of a scalar function of n variables $f(x_1, x_2, \ldots, x_n)$ is the column vector $\dfrac{\partial f}{\partial x}$ defined by

$$\frac{\partial f}{\partial x} \triangleq \begin{bmatrix} \dfrac{\partial f}{\partial x_1} \\[2mm] \dfrac{\partial f}{\partial x_2} \\ \cdot \\ \cdot \\ \cdot \\ \dfrac{\partial f}{\partial x_n} \end{bmatrix} \tag{1.35}$$

If $f(x)$ is linear in x, that is, $f(x) = d'x$ where d is a constant n-dimensional column vector, then the gradient is a constant vector

$$\frac{\partial f}{\partial x} = d$$

If $f(x)$ is a quadratic form, i.e., $f(x) = x'Ax$ then

$$\frac{\partial f}{\partial x} = Ax + A'x$$

* Any nth order polynomial $\lambda^n - \alpha_1 \lambda^{n-1} + \cdots + \alpha_n = 0$ has the property that

$$\alpha_1 = \sum_{i=1}^{n} \lambda_i$$

where λ_i are the roots of the polynomial.

1.13   The Cayley-Hamilton Theorem

The Cayley-Hamilton theorem states that every matrix satisfies its own characteristic equation. (More specifically, the matrix satisfies the "matrix polynomial version" of the characteristic equation.)

Proof: Since for a value of λ not equal to any of the eigenvalues of A, we have

$$|A - \lambda I| \neq 0$$

It follows that the matrix $A - \lambda I$ is nonsingular. Let

$$[A - \lambda I]^{-1} = \frac{B(\lambda)}{g(\lambda)} \tag{1.36}$$

or

$$[A - \lambda I]B(\lambda) = g(\lambda)I \tag{1.37}$$

where

$$g(\lambda) = |A - \lambda I| = (-1)^n \lambda^n + \alpha_1 \lambda^{n-1} + \cdots \alpha_n \tag{1.38}$$

is the characteristic polynomial of A. Furthermore, $B(\lambda)$, according to equation (1.4) is a matrix with elements which are polynomials in λ with order not larger than λ^{n-1}. Consequently, we can write

$$B(\lambda) = \lambda^{n-1}B_{n-1} + \lambda^{n-2}B_{n-2} + \cdots + B_0 \tag{1.39}$$

where B_i is an $n \times n$ matrix independent of λ. Substituting (1.38) and (1.39) into (1.37) and equating coefficients of λ yields

$$-B_{n-1} = (-1)^n I$$

$$AB_{n-1} - B_{n-2} = \alpha_1$$

$$AB_{n-2} - B_{n-3} = \alpha_2 \tag{1.40}$$

$$\cdot \quad \quad \cdot \quad \quad \cdot$$
$$\cdot \quad \quad \cdot \quad \quad \cdot$$
$$\cdot \quad \quad \cdot \quad \quad \cdot$$

$$AB_1 \quad - B_0 \quad = \alpha_n$$

Therefore

$$B_{n-1} = -(-1)^n I$$

$$B_{n-2} = -(-1)^n A - \alpha_1$$

$$B_{n-3} = -(-1)^n A^2 - \alpha_1 A - \alpha_2 \tag{1.41}$$

$$\cdot \quad \quad \cdot \quad \quad \cdot$$
$$\cdot \quad \quad \cdot \quad \quad \cdot$$
$$\cdot \quad \quad \cdot \quad \quad \cdot$$

that is, every B_i is a polynomial in A with scalar coefficients. It follows that B_iA is a polynomial in A and hence $B_iA = AB_i$. Hence the identity (1.37) is valid when $\lambda = A$. Therefore

$$g(A) = 0 \qquad (1.42)$$

which is the desired result.

▶ **example**

The matrix

$$A = \begin{bmatrix} 2 & 1 \\ 3 & 4 \end{bmatrix}$$

has the characteristic equation

$$\lambda^2 - 6\lambda + 5 = 0$$

Therefore, by the Cayley-Hamilton theorem,

▶ $$A^2 - 6A + 5I = 0$$

DIFFERENTIAL AND DIFFERENCE EQUATIONS

1.14 General Remarks

A differential equation is a relationship among a function $x(t)$, its derivatives $\dot{x}(t)$, $\ddot{x}(t)$, etc., and the independent variable t. The order of a differential equation is the order of the highest derivative existing in the relationship. Therefore, if a function is given by $x(t)$, a first order differential equation is a relationship between $x(t)$, $\dot{x}(t)$, and t which in general can be written implicitly as

$$g[x(t), \dot{x}(t), t] = 0$$

It is usually the case, in engineering practice, that the above implicit equation can be solved for $\dot{x}(t)$ as an explicit function of $x(t)$ and t:

$$\dot{x}(t) = f[x(t), t] \qquad (1.43)$$

In more general terms, the function f not only depends explicitly on t and $x(t)$, but may also depend on t through another function (usually known), say, $u(t)$. For example,

$$\dot{x}(t) = f[x(t), t, u(t)] \qquad (1.44)$$

The function $u(t)$ usually represents an "input" or "control" which is applied to the dynamic system represented by the differential equation (1.44). By definition, a solution to the differential equation (1.44) associated with an input $u(t)$ and an initial condition $x(t_0) = x_0$ is a function of time depending on $u(t)$ and x_0 denoted here by

$$\varphi[t, t_0, x_0, u(t)] \tag{1.45}$$

satisfying the following conditions:*

$$\dot{\varphi}[t, t_0, x_0, u(t)] = f[\varphi(t, t_0, x_0, u(t)), t, u(t)] \tag{1.46}$$

$$\varphi[t_0, t_0, x_0, u(t)] = x_0 \tag{1.47}$$

The reason for specifying the initial condition $x(t_0) = x_0$ is that in its absence most meaningful systems (1.44) will yield an infinite number of functions φ which satisfy the condition (1.46). For simplicity in notation we shall use $x(t)$ in place of $\varphi[t, t_0, x_0, u(t)]$ when this causes no ambiguity (i.e., when a specific initial time t_0, initial state x_0, and input $u(t)$ are understood).

The functions $x(t)$, $f[x(t), t, u(t)]$, and $u(t)$ in general are vector-valued. The components of $x(t)$ are called "state variables" and $x(t)$ the "state vector" (more simply the "state"). The state variables have the property that at any time, such as $t = t_0$, they represent the minimum number of variables which uniquely† define the solution $x(t)$ for $t \geq t_0$ if $u(t)$ for $t \geq t_0$ is known.

A vector-valued differential equation given by (1.43) may correspond to one or more scalar differential equations of order larger than 1, and vice versa. For example, starting with the following differential equation

$$\begin{bmatrix} \dot{x}_1(t) \\ \dot{x}_2(t) \end{bmatrix} = \begin{bmatrix} x_2(t) \\ x_1^3(t) + x_2(t) \end{bmatrix} \tag{1.48}$$

we can use the fact that $\dot{x}_1(t) = x_2(t)$ to write

$$\ddot{x}_1(t) = \dot{x}_2(t)$$

Substituting for $\dot{x}_2(t)$ from (1.48) yields the *second order, scalar* differential equation

$$\ddot{x}_1(t) - \dot{x}_1(t) - x_1^3(t) = 0 \tag{1.49}$$

To see that (1.49) is equivalent to (1.48) we can simply introduce a new variable $x_2(t)$ by defining

$$\dot{x}_1(t) \triangleq x_2(t) \tag{1.50}$$

Differentiating $x_2(t)$ and using (1.49) it follows that

$$\dot{x}_2(t) = x_1^3(t) + x_2(t) \tag{1.51}$$

Hence starting with equation (1.49) we can derive equation (1.48).

* By definition, the term $u(t)$ in the argument of $\varphi[t, t_0, x_0, u(t)]$ denotes the dependence of φ on $u(\tau)$, $t_0 \leq \tau \leq t$.
† If a unique solution exists.

It is important to observe that (1.48) is not a unique representation (in the form of a first order vector differential equation) of (1.49) since the choice of $x_2(t)$ is arbitrary. For example, if we choose $x_2(t)$ by

$$\dot{x}_1(t) = x_1(t) + x_2(t) \tag{1.52}$$

then substituting into (1.49) and solving for $\dot{x}_2(t)$ it follows that

$$\dot{x}_2(t) = x_1{}^3(t) \tag{1.53}$$

It is seen that while equation (1.48) appears to be different from equations (1.52) and (1.53), both are equivalent to (1.49).

When the variable t is discrete, that is, the functions x, u, and f are represented by $x(k)$, $u(k)$, and $f[x(k), k, u(k)]$ where k is a discrete variable (such as $k = 1, 2, 3, \ldots$) a *difference equation* is defined by

$$x(k + 1) = f[x(k), k, u(k)] \tag{1.54}$$

to which the concepts of solution and state variables discussed in terms of differential equations are directly applicable. Difference equations have dynamic properties which are quite similar to those of differential equations.

In this book, wherever differential or difference equations are used, it is assumed that they possess a solution and that the solution is unique.* Actually, since the difference equation (1.54) explicitly specifies how to compute $x(k + 1)$ from $x(k)$, k, and $u(k)$, the solution always exists and is unique.

1.15 Linearity

If the vector-valued functions f in (1.44) or (1.54) are linear functions of x and u, the differential or difference equations are called "linear" and can be written

$$\dot{x}(t) = A(t)x(t) + B(t)u(t) \tag{1.55}$$

or

$$x(k + 1) = A(k)x(k) + B(k)u(k) \tag{1.56}$$

where x and u are n and r vectors, and A and B are $n \times n$ and $n \times r$ matrices respectively.

Linear differential (or difference) equations have the important property that the solution (or "response") to a sum of different inputs and/or a sum of different initial conditions is equal to the sum of the individual solutions. This is called the principle of superposition; that is, if $x^1(t)$ is the solution corresponding to $u^1(t)$ and $x^1(t_0) = x_0{}^1$ and $x^2(t)$ corresponds to $u^2(t)$ and

---

* For conditions establishing existence and uniqueness of solutions, cf. Coddington and Levinson, *Theory of Ordinary Differential Equations*.

$x^2(t_0) = x_0^2$ then $x^3(t)$ will be the solution corresponding to $u^1(t) + u^2(t)$ and initial condition $x_0^1 + x_0^2$ where $x^3(t) = x^1(t) + x^2(t)$. This is verified as follows: The solutions $x^1(t)$ and $x^2(t)$ by definition satisfy (1.55). Hence

$$\dot{x}^1(t) = A(t)x^1(t) + B(t)u^1(t) \qquad x^1(t_0) = x_0^1 \tag{1.57}$$

$$\dot{x}^2(t) = A(t)x^2(t) + B(t)u^2(t) \qquad x^2(t_0) = x_0^2 \tag{1.58}$$

Adding these two equations yields

$$\dot{x}^1(t) + \dot{x}^2(t) = A(t)[x^1(t) + x^2(t)] + B(t)[u^1(t) + u^2(t)] \tag{1.59}$$

Therefore, by definition of a solution, $x^3(t) = x^1(t) + x^2(t)$ is the solution corresponding to $u^1(t) + u^2(t)$ and $x^3(t_0) = x^1(t_0) + x^2(t_0)$.

Identical results can be obtained for the linear difference equation (1.56). The principle of superposition of solutions may hold for certain differential (difference) equations of more general form than (1.55) or (1.56), but these cases will not concern us in this book.

1.16 Autonomous Systems

If the function f is neither a function of $u(t)$ nor an explicit function of t, the differential or difference equation is called "autonomous." In the case of linear equations, this implies that the matrix A is a constant and that $u(t) = 0$ (or $B = 0$). If only the explicit dependence of f on t is not present, the equations are called time-invariant. It should be noted that, for any given input $u(t)$, the right hand side of (1.44) can be written in overall form as $f(x, t)$, where the dependence of $f(x, t)$ on t may be either the result of an independent forcing function $u(t)$, or the result of the original dependence of $f[x(t), t, u(t)]$ explicitly on t, or both. Questions of existence and uniqueness of solutions of (1.44) must always be answered in terms of the function $f(x, t)$ representing the *overall* dependence of (1.44) on x and t.

1.17 Homogeneous Solution

The solution of a differential or difference equation when the forcing function is set equal to zero is called the homogeneous solution. In other words, the homogeneous solution is the response to initial conditions only. It turns out in Section 1.18 that the forced, or inhomogeneous, solution to a differential equation can be conveniently expressed in terms of a particular homogeneous solution, called the transition matrix, which will now be introduced.

Consider the vector differential equation

$$\dot{x}(t) = A(t)x(t)$$
$$x(t_0) = x_0 \tag{1.60}$$

where $x(t)$ is an n-vector and $A(t)$ is an $n \times n$ matrix. Since there is no input $u(t)$ in (1.60), the solution $x(t)$ is certainly a homogeneous solution. Let us now consider a set of n differential equations of the form (1.60) where $A(t)$ remains the same for all the differential equations in the set, but where the initial condition vectors x_0 are (possibly) all different:

$$\dot{x}^1 = A(t)x^1 \quad \dot{x}^2 = A(t)x^2 \quad \cdots \quad \dot{x}^n = A(t)x^n \tag{1.61}$$

$$x^1(t_0) = x_0{}^1 \quad x^2(t_0) = x_0{}^2 \quad x^n(t_0) = x_0{}^n \tag{1.62}$$

Here, superscripts have been used to index the n differential equations. We can express these equations by defining the $n \times n$ matrix $X(t)$ as

$$X(t) = [x^1(t), x^2(t), \ldots, x^n(t)] \tag{1.63}$$

where each column of $X(t)$ is a solution to one of the differential equations in the set. Then, the $n \times n$ matrix function $X(t)$ satisfies the $n \times n$ *matrix* differential equation

$$\dot{X}(t) = A(t)X(t) \tag{1.64}$$

with, now, an $n \times n$ *matrix* initial condition

$$X(t_0) = [x_0{}^1, x_0{}^2, \ldots, x_0{}^n] \tag{1.65}$$

Let us consider a specific choice of the $n \times n$ initial condition $X(t_0)$, namely

$$X(t_0) = I \tag{1.66}$$

and the corresponding homogeneous solution which we shall denote by $X(t, t_0)$. This rather special $n \times n$ matrix solution of (1.64) is guaranteed to exist, and is called the "fundamental* solution matrix" or "transition matrix" of the original $n \times 1$ differential equation (1.60). By definition, $X(t, t_0)$ satisfies

$$\dot{X}(t, t_0) = A(t)X(t, t_0) \tag{1.67}$$

$$X(t_0, t_0) = I \tag{1.68}$$

We shall now show that the (homogeneous) solution of (1.60) can be written in terms of the transition matrix: Multiplying both sides of (1.67) by $x(t_0)$, the vector initial condition of (1.60), yields

$$\dot{X}(t, t_0)x(t_0) = A(t)X(t, t_0)x(t_0) \tag{1.69}$$

Denoting

$$y(t) = X(t, t_0)x(t_0) \tag{1.70}$$

* In mathematical terminology, any $n \times n$ matrix solution of (1.64) having linearly independent columns is called a "fundamental solution matrix," whether or not it has initial condition $X(t_0) = I$. The engineering terminology "transition matrix" is reserved for that unique fundamental solution matrix having the initial condition $X(t_0) = I$.

and substituting into (1.69) a differential equation identical with (1.60) results. Furthermore, from (1.70) and since $X(t_0, t_0) = I$, we obtain

$$y(t_0) = x(t_0) \tag{1.71}$$

Consequently, $y(t)$ is equal to the solution of (1.60), i.e.,

$$x(t) = X(t, t_0)x(t_0) = X(t, t_0)x_0 \tag{1.72}$$

In other words, the homogeneous solution of the vector differential equation (1.60) may be expressed in terms of its transition matrix.

An important property of the transition matrix is that it is always a non-singular matrix. Let

$$X(t, t_0) = [c^1 c^2, \ldots, c^n] \tag{1.73}$$

where c^i is the ith column of $X(t, t_0)$. Choose n initial conditions as follows

$$x^1(t_0) = \begin{bmatrix} 1 \\ 0 \\ \cdot \\ \cdot \\ \cdot \end{bmatrix}$$

$$\tag{1.74}$$

$$x^2(t_0) = \begin{bmatrix} 0 \\ 1 \\ 0 \\ \cdot \\ \cdot \\ \cdot \end{bmatrix}$$

and denote the corresponding solutions of (1.60) by $x^1(t), \ldots, x^n(t)$. From (1.72) these are the columns of the transition matrix $X(t, t_0)$, i.e.,

$$x^i(t) = c^i, \qquad i = 1, \ldots, n \tag{1.75}$$

Consequently, in order to prove that $X(t, t_0)$ is nonsingular it is necessary and sufficient that the n vectors $x^i(t)$ be linearly independent. Let us assume they are not. Then there exist n constants $\alpha_1, \ldots, \alpha_n$ such that

$$\alpha_1 x^1(t) + \alpha_2 x^2(t) + \cdots + \alpha_n x^n(t) = 0 \tag{1.76}$$

Therefore, we should also have

$$\alpha_1 x^1(t_0) + \alpha_2 x^2(t_0) + \cdots + \alpha_n x^n(t_0) = 0 \tag{1.77}$$

From (1.74) this is a contradiction.

The matrix $X(t, t_0)$ is called the transition matrix because it causes the initial value of x at t_0, namely, $x(t_0)$ to *move* to $x(t)$ at t. We may briefly derive some useful properties of the transition matrix. From (1.72) we have

$$x(t_1) = X(t_1, t_0)x(t_0) \tag{1.78}$$

and

$$x(t_2) = X(t_2, t_0)x(t_0) \tag{1.79}$$

If the initial condition is specified as $x(t_1)$, the solution at t_2 will become

$$x(t_2) = X(t_2, t_1)x(t_1) \tag{1.80}$$

Substituting for $x(t_2)$ and $x(t_1)$ from (1.78) and (1.79) yields

$$X(t_2, t_0)x(t_0) = X(t_2, t_1)X(t_1, t_0)x(t_0) \tag{1.81}$$

Since (1.81) is valid for an arbitrary $x(t_0)$, we have proved the "transitive" relationship

$$X(t_2, t_0) = X(t_2, t_1)X(t_1, t_0) \tag{1.82}$$

which is valid for any t_0, t_1, and t_2. If we now choose $t_2 = t_0$ it follows that

$$X(t_0, t_1)X(t_1, t_0) = X(t_0, t_0) = I \tag{1.83}$$

$$X^{-1}(t_0, t_1) = X(t_1, t_0) \tag{1.84}$$

Hence, the inverse of the transition matrix from t_1 to t_0 is obtained by simply reversing the arguments in the transition matrix from t_0 to t_1.

▶ **example**
Consider the homogeneous differential equation

$$\dot{x}_1(t) = \frac{-1}{1 - 2t} [x_1(t) - x_2(t)]$$

$$\dot{x}_1(t) = \frac{1}{1 - 2t} [x_1(t) - x_2(t)]$$

The transition matrix is given by

$$X(t, t_0) = \begin{bmatrix} \dfrac{1 - t - t_0}{1 - 2t_0} & \dfrac{t - t_0}{1 - 2t_0} \\ \dfrac{t - t_0}{1 - 2t_0} & \dfrac{1 - t - t_0}{1 - 2t_0} \end{bmatrix} \tag{1.85}$$

This is verified by substituting into (1.67) where

$$A(t) = \frac{1}{1 - 2t} \begin{bmatrix} -1 & +1 \\ 1 & -1 \end{bmatrix}$$

Furthermore, by directly inverting $X(t, t_0)$ in (1.85), we obtain

$$X^{-1}(t, t_0) = \begin{bmatrix} \dfrac{1 - t_0 - t}{1 - 2t} & \dfrac{t_0 - t}{1 - 2t} \\[3mm] \dfrac{t_0 - t}{1 - 2t} & \dfrac{1 - t_0 - t}{1 - 2t} \end{bmatrix}$$

▶

In general, analytical derivations of transition matrices for various differential equations are not simple and one may have to resort to computational procedures with the aid of a digital or analog computer.

In the sequel, we will always let $t_0 \triangleq 0$ without loss of generality, and agree to represent $X(t, t_0)$ by $X(t)$, so that $X(0) = I$.

1.18 The Inhomogeneous Solution

We are now in position to obtain a solution to the inhomogeneous differential equation

$$\dot{x}(t) = A(t)x(t) + B(t)u(t) \tag{1.86}$$

$$x(0) = x_0$$

Using the method of undetermined multipliers (Lagrange) let

$$x(t) = X(t)z(t) \tag{1.87}$$

where $X(t)$ is the transition matrix. If this is the right form for the solution it should satisfy (1.86) as an identity, i.e.,

$$\dot{X}(t)z(t) + X(t)\dot{z}(t) = A(t)X(t)z(t) + B(t)u(t) \tag{1.88}$$

Since

$$\dot{X}(t) = A(t)X(t)$$

and consequently

$$\dot{X}(t)z(t) = A(t)X(t)z(t)$$

It follows that

$$X(t)\dot{z}(t) = B(t)u(t) \tag{1.89}$$

Solving (1.89) for $z(t)$ yields

$$z(t) = z(0) + \int_0^t X^{-1}(s)B(s)u(s)\, ds \tag{1.90}$$

where $z(0)$ is the initial condition. From (1.87) and recalling that $X(0) = I$ we have

$$z(0) = x(0) \tag{1.91}$$

Substituting (1.90) and (1.91) into (1.87) yields

$$x(t) = X(t)\left(x(0) + \int_0^t X^{-1}(s)B(s)u(s)\, ds \right) \tag{1.92}$$

▶ **example**

Let $x(t)$ be a scalar function governed by the differential equation

$$\dot{x}(t) = (1 + 2t)x(t) + 1$$
$$x(0) = 2$$

The transition matrix is the solution of the equation

$$\dot{X}(t) = (1 + 2t)X(t)$$

Since $X(t)$ is scalar here, we can write

$$\frac{dX(t)}{X(t)} = (1 + 2t)\, dt$$

Integrating both sides yields

$$X(t) = e^{t+t^2} + c$$

Since

$$X(0) = I$$

it follows that

$$X(t) = e^{t+t^2}$$

Substituting into (1.92) yields

$$x(t) = e^{t+t^2}\left(2 + \int_0^t e^{-(s+s^2)}\, ds\right)$$

1.19 Linear Time-Invariant Systems

A linear time-invariant vector differential equation is given by

$$\dot{x}(t) = Ax(t) + Bu(t) \qquad (1.93)$$

where A and B are constant matrices of dimensions $n \times n$ and $n \times r$, respectively. Let us define the matrix exponential e^{At} as the following matrix power series:

$$e^{At} = I + \frac{t}{1!}A + \frac{t^2}{2!}A^2 + \frac{t^3}{3!}A^3 \qquad (1.94)$$

We can formally determine the derivative $\frac{d}{dt}e^{At}$ (from the series given by (1.94)) as

$$\frac{d}{dt}e^{At} = 0 + A + \frac{2t}{2!}A^2 + \frac{3t^2}{3!}A^3 + \cdots \qquad (1.95)$$

Factoring A in the right hand side of (1.95) and using (1.94) yields

$$\frac{d}{dt}e^{At} = Ae^{At} \qquad (1.96)$$

Therefore, the matrix e^{At} as defined by (1.94) will satisfy the matrix differential equation

$$\dot{X}(t) = AX(t)$$

and consequently the transition matrix when A is a constant matrix is given by

$$X(t) = e^{At} \tag{1.97}$$

From the property of the transition matrix we have

$$X^{-1}(t) = e^{-At} \tag{1.98}$$

Finally, the solution to the inhomogeneous equation (1.93) is obtained from (1.92).

$$x(t) = e^{At}x(0) + \int_0^t e^{A(t-s)}Bu(s)\,ds \tag{1.99}$$

▶ **example**

Let a differential equation be given by

$$\dot{x}_1(t) = x_2(t)$$
$$\dot{x}_2(t) = -x_1(t) + t$$
$$x_1(0) = x_2(0) = 0$$

Hence

$$A = \begin{bmatrix} 0 & 1 \\ -1 & 0 \end{bmatrix}$$

$$B = \begin{bmatrix} 0 \\ 1 \end{bmatrix}$$

$$u(t) = t$$

Then from (1.94)

$$e^{At} = I + \begin{bmatrix} 0 & 1 \\ -1 & 0 \end{bmatrix} t + \begin{bmatrix} -1 & 0 \\ 0 & -1 \end{bmatrix} \frac{t^2}{2!}$$

$$+ \begin{bmatrix} 0 & -1 \\ 1 & 0 \end{bmatrix} \frac{t^3}{3!} + \begin{bmatrix} 1 & 0 \\ 0 & 1 \end{bmatrix} \frac{t^4}{4!} + \cdots$$

or

$$e^{At} = \begin{bmatrix} 1 - \dfrac{t^2}{2} + \dfrac{t^4}{4!} - \cdots & t - \dfrac{t^3}{3!} + \cdots \\[2ex] -t + \dfrac{t^3}{3!} - \cdots & 1 - \dfrac{t^2}{2!} + \dfrac{t^4}{4!} - \cdots \end{bmatrix}$$

The elements of this matrix are recognized as the series expansions for $\sin t$ and $\cos t$, hence

$$e^{At} = \begin{bmatrix} \cos t & \sin t \\ -\sin t & \cos t \end{bmatrix}$$

Substituting into (1.99) yields

$$x(t) = \int_0^t \begin{bmatrix} \cos(t-s) & \sin(t-s) \\ -\sin(t-s) & \cos(t-s) \end{bmatrix} \begin{bmatrix} 0 \\ s \end{bmatrix} ds$$

and finally

▶
$$x(t) = \begin{bmatrix} t - \sin t \\ 1 - \cos t \end{bmatrix}$$

1.20 Impulse Response

In a linear system given by

$$\dot{x}(t) = A(t)x(t) + B(t)u(t) \tag{1.100}$$
$$x(0) = 0$$

let the input $u(t)$ be scalar (i.e., B is an $n \times 1$ matrix). Consider the scalar output $y(t)$ given by

$$y(t) = C(t)x(t) \tag{1.101}$$

where C is a $1 \times n$ row vector. In such a linear system the variables $u(t)$ and $y(t)$ are related to each other by the integral

$$y(t) = \int_0^t h(t, s)u(s)\, ds \tag{1.102}$$

where $h(t, s)$ is called the system impulse response. On the other hand, if we let $x(0) = 0$ in (1.92) and premultiply both sides by C, it follows that

$$y(t) = \int_0^t C(t)X(t)X^{-1}(s)B(s)u(s)\, ds \tag{1.103}$$

Since (1.102) and (1.103) are equivalent representations, it follows that $h(t, s)$ may be expressed as

$$h(t, s) = C(t)X(t)X^{-1}(s)B(s) \tag{1.104}$$

where for the special case of constant-coefficient systems, we have

$$h(t, s) = h(t - s) = Ce^{A(t-s)}B \tag{1.105}$$

or

$$h(t) = Ce^{At}B$$

In general, if C is an $m \times n$ and B an $n \times r$ matrix, the term $Ce^{At}B$ represents an $m \times r$ matrix consisting of elements each of which is the impulse response corresponding to a pair of input-output variables.

▶ **example**

Consider the system

$$\dot{x}_1(t) = x_2(t)$$
$$\dot{x}_2(t) = -x_1(t) + u(t)$$

Let us denote $u(t)$ as the scalar input and $x_1(t)$ as the scalar output. Then by (1.101) we have

$$C(t) = [1 \quad 0], \quad B(t) = \begin{bmatrix} 0 \\ 1 \end{bmatrix}$$

From the preceding example we have

$$e^{At} = \begin{bmatrix} \cos t & \sin t \\ -\sin t & \cos t \end{bmatrix}$$

The impulse response, i.e., the solution $x_1(t)$ in response to $u(t) = \delta(t)$ is now given by (1.105):

$$▶ \qquad h(t) = [1 \quad 0]e^{At}\begin{bmatrix} 0 \\ 1 \end{bmatrix} = \sin t$$

1.21 Evaluation of Transition Matrix

It should be evident by now that the major difficulty in solving linear differential equations is the determination of their transition matrices. In general, there exists no simple procedure to evaluate $X(t)$, and in practice it can only be evaluated by means of computational methods. The following infinite series expansion of $X(t)$ was used by Kinariwala* and can be of some use in this regard

$$X(t) = I + \int_0^t A(\tau_1)\, d\tau_1 + \int_0^t A(\tau_2) \int_0^{\tau_2} A(\tau_1)\, d\tau_1\, d\tau_2 + \cdots$$

Of course, the series has to be truncated, and hence we can only get an approximation to $X(t)$.

A direct approach is to solve the matrix equation

$$\dot{X}(t) = A(t)X(t) \qquad\qquad (1.106)$$
$$X(0) = I$$

* R. K. Kinariwala, "Analysis of Time Varying Networks," *IRE International Convention Record*, 1961, Pt.4.

directly by a computer. We can also approximate the terms of $X(t)$ by polynomials (usually in t) and evaluate their coefficients by satisfying (1.106) in some approximate manner.

Where A is constant, the problem is much simpler. We can use the expansion of e^{At} given by (1.94) and either truncate the series or try to recognize some trend by evaluating the first few terms in the series. The latter procedure was used in the example in Section 1.19.

The method of Laplace transformation can be used to evaluate e^{At} analytically. If $f(t)$ and $\bar{f}(s)$ are a Laplace transform pair we have

$$\bar{f}(s) = \int_0^\infty f(t)e^{-st}\, dt \tag{1.107}$$

The Laplace transform operation can be applied to $f(t)$ when it is a matrix since it only involves multiplication by a scalar e^{-st} and integration. Taking the Laplace transform of both sides of the matrix equation

$$\dot{X}(t) = AX(t)$$
$$X(0) = I$$

and recalling the result associated with the Laplace transform of the derivative of a function, namely,

$$\mathscr{L}\dot{X}(t) = s\bar{X}(s) - X(0)$$

it follows that

$$\bar{X}(s)[sI - A] = I$$

or

$$\bar{X}(s) = [sI - A]^{-1} \tag{1.108}$$

The desired transition matrix is then obtained by finding the inverse Laplace transform of $\bar{X}(s)$. The inverse Laplace transformation is applied to each element of $\bar{X}(s)$ to obtain the elements of the transition matrix $X(t)$.

▶ **example**

Let us find the transition matrix for the example in Section 1.19 where

$$A = \begin{bmatrix} 0 & 1 \\ -1 & 0 \end{bmatrix}$$

substituting into (1.108) yields

$$\bar{X}(s) = \begin{bmatrix} s & -1 \\ 1 & s \end{bmatrix}^{-1} = \begin{bmatrix} \dfrac{s}{s^2+1} & \dfrac{1}{s^2+1} \\[2ex] \dfrac{-1}{s^2+1} & \dfrac{s}{s^2+1} \end{bmatrix}$$

which has the inverse transform given by

▶
$$X(t) = \begin{bmatrix} \cos t & \sin t \\ -\sin t & \cos t \end{bmatrix}$$

1.22   Linear Difference Equations

The solution to a linear vector difference equation of the form

$$x(k + 1) = A(k)x(k) + B(k)u(k)$$
$$x(0) = x_0$$

$$(1.109)$$

can be constructed by direct matrix multiplication as follows

$x(1) = A(0)x(0) + B(0)u(0)$

$x(2) = A(1)x(1) + B(1)u(1) = A(1)A(0)x(0) + A(1)B(0)u(0) + B(1)u(1)$

$x(3) = A(2)x(2) + B(2)u(2) = A(2)A(1)A(0)\,x(0) + A(2)A(1)B(0)u(0)$
$$+ A(2)B(1)u(1) + B(2)u(2)$$

.
.
.

or in the compact (but not very useful) form

$$x(k) = \left[\prod_{i=k-1}^{0} A(i) \right] x(0) + \sum_{i=0}^{k-1} \left[\begin{array}{l} \prod\limits_{j=k-1}^{i+1} A(j) \; ; \; i + 1 \leqslant k - 1 \\[2mm] I \; ; \; (i + 1) > k - 1 \end{array} \right] B(i)u(i) \quad (1.110)$$

For the constant-coefficient system, (1.110) becomes

$$x(k) = A^{k-1}x(0) + \sum_{i=0}^{k-1} A^{k-(i+1)}Bu(i) \qquad (1.111)$$

If $A\,(j)j = 0, 1, \ldots, k - 1$ are nonsingular matrices, (1.110) can also be written in the following form

$$x(k) = \left[\prod_{i=k-1}^{0} A(i) \right] \left[x(0) + \sum_{i=0}^{k-1} \left(\prod_{j=i}^{0} A(j) \right)^{-1} B(i)u(i) \right] \qquad (1.112)$$

Comparing (1.112) and (1.92) it is evident that the term

$$\prod_{j=i}^{0} A(j)$$

plays the role of transition matrix $X(t)$ and it can be easily verified that it possesses properties similar to those of $X(t)$. When $A(k)$ is singular, this gives the "forward" transitions of the system (1.109), but the "reverse" transitions are undefined.

▶ **example**

Let

$$A(k) = \begin{bmatrix} 1 & 1 \\ 0 & 1 \end{bmatrix}$$

$$B(k) = \begin{bmatrix} 0 \\ 1 \end{bmatrix}$$

$$u(k) = k$$

$$x(0) = 0$$

Then

$$A^k = \begin{bmatrix} 1 & k \\ 0 & 1 \end{bmatrix}$$

and from (1.111)

$$x(k) = \sum_{i=0}^{k-1} \begin{bmatrix} 1 & k - (1 + i) \\ 0 & 1 \end{bmatrix} \begin{bmatrix} 0 \\ 1 \end{bmatrix} i$$

Hence

▶

$$x(k) = \frac{k(k-1)}{3} \begin{bmatrix} \dfrac{5k-4}{3} \\ 1 \end{bmatrix}$$

ELEMENTS OF
RANDOM PROCESSES

1.23 Events and Probabilities

Let us define a statistical experiment as a physical or theoretical process and call the outcome of the experiment an "event." An example of a statistical experiment is throwing a die. The events, or outcomes, consist of a particular face of the die turning up after the throw. Events are required to have certain properties: If α is an event, then the *non*-occurrence of α is also an event. If α and β are two events, then the joint occurrence of α *and* β is an event. Furthermore, the occurrence of either α or β or both (considered as one overall possibility which could occur) is an event. Among the possible outcomes, there is always a *certain* event, or event which always occurs. If the occurrence of an event α is certain, its probability $p(\alpha)$ is defined as unity (such as the event consisting of any of all possible outcomes). For any arbitrary event β we have $0 \leq p(\beta) \leq 1$.

The events α and β are called disjoint (mutually exclusive) if occurrence of one implies that the other cannot occur. If $\alpha_1, \alpha_2, \ldots, \alpha_n$ are a set of disjoint events, then

$$p(\alpha_1 \text{ or } \alpha_2 \text{ or } \alpha_3, \ldots \text{ or } \alpha_n) = \sum_{i=1}^{n} p(\alpha_i) \qquad (1.113)$$

If it is possible to decompose the "certain" event into disjoint events $\alpha_1, \ldots, \alpha_n$, then by definition

$$\sum_{i=1}^{n} p(\alpha_i) = 1 \qquad (1.114)$$

In the example of throwing a die the outcomes are any of the possible six faces. For a perfectly symmetrical die the probability associated with any of the six possible outcomes is $1/6$. These events are mutually exclusive, and, since the certain event is that one of the six faces will come up, the sum of probabilities associated with all outcomes, namely, faces $1, 2, \ldots, 6$, is unity. It is possible for the outcome of throwing a die to be a more complicated kind of event than the mere showing of a single face. We can see this by considering the event that either the 6-face, or the 4-face, or the 1-face occurs. Clearly, if we throw the die and see that the 1-face comes up, it is obvious that the above-described event occurs. At this point it seems desirable to define the showing of a simple face as an "elementary event," one of which is certain to occur. Then we can define "more complicated events" as logical sums, products, and complements of the elementary events.

The total number of possible events for any experiment may be finite, as in the example of throwing a dice, or infinite. An example of the latter is the position x of a randomly moving particle. Since the position of the particle is a continuous function, the probability density $p(x)$ is defined for *any* value of x. The notions introduced above and in what follows apply to continuous probability densities as well.

1.24 Joint, Marginal, and Conditional Probabilities

In the preceding section we were concerned only with the outcomes of a single experiment. In general, we might be concerned with outcomes of several experiments (which usually have something in common). For example, we may be interested in the noise amplitude at several instants of time. The probabilities relating to such events are referred to as joint probability density functions. Suppose the outcomes of one experiment are $\alpha_1, \ldots, \alpha_K$, and the outcomes of another experiment are $\beta_1, \ldots, \beta_L$. The joint probability of the kth and lth events of the two experiments respectively is denoted by

$$p(\alpha_k, \beta_l) \tag{1.115}$$

Throwing of two dice is also an experiment. An outcome or event then is the showing of any pair of faces. This event may be regarded as either a joint event of two experiments (each consisting of throwing one die) or an event of one experiment (consisting of throwing two dice at once). As a joint event, it possesses a joint probability density with a value symbolized as $p(2, 3)$, which represents the (joint) probability of appearance of face two of dice number one and face 3 of dice number 2. It is easily seen that the joint density for any pair of faces has a value equal to $1/36$. Since any two experiments can be considered as one with the corresponding *joint* event (α_k, β_l) then

$$0 \le p(\alpha_k, \beta_l) \le 1 \tag{1.116}$$

and if α's and β's are disjoint we have

$$\sum_{k=1}^{K} \sum_{l=1}^{L} p(\alpha_k, \beta_l) = 1 \tag{1.117}$$

From the definition of joint probability we can observe that

$$p(\alpha_k) = \sum_{l=1}^{L} p(\alpha_k, \beta_l) \tag{1.118}$$

In (1.118) $p(\alpha_k)$ is called the "marginal" density on α_k, to imply that it has been derived from a joint density by the indicated summation.

Another interesting probability density is that of β_l when we know that α_k has occurred. This is denoted by the conditional probability

$$p(\beta_l \mid \alpha_k) \tag{1.119}$$

and is read as the probability of β_l given α_k. The conditional probability (1.119) is defined implicitly in terms of the joint density $p(\alpha_k, \beta_l)$ and the marginal density $p(\alpha_k)$ by $p(\beta_l \mid \alpha_k)p(\alpha_k) \triangleq p(\alpha_k, \beta_l)$. If $p(\alpha_k) \ne 0$, the conditional density can be written explicitly as

$$p(\beta_l \mid \alpha_k) = \frac{p(\alpha_k, \beta_l)}{p(\alpha_k)} \tag{1.120}$$

The important relationship given in (1.120) is called *Bayes' Formula*. Since a conditional probability density is itself a probability density, it has the properties

$$0 \le p(\beta_l \mid \alpha_k) \le 1 \quad \text{for each } \alpha_k \tag{1.121}$$

and

$$\sum_{l=1}^{L} p(\beta_l \mid \alpha_k) = 1 \quad \text{for each } \alpha_k \tag{1.122}$$

and furthermore

$$p(\beta_l \mid \alpha_k) \geq p(\alpha_k, \beta_l) \tag{1.123}$$

which follows from (1.120) by the fact that $p(\alpha_k) \leq 1$.

1.25 Statistical Independence

The probability $p(\beta_l \mid \alpha_k)$ is the probability of occurrence of β_l conditioned on the fact that α_k has occurred. It is possible that the occurrence of the event α_k has no influence on the occurrence of event β_l, i.e.,

$$p(\beta_l \mid \alpha_k) = p(\beta_l) \tag{1.124}$$

Hence, from (1.120)

$$p(\alpha_k, \beta_l) = p(\alpha_k)p(\beta_l) \tag{1.125}$$

Alternatively, equation (1.120) can be written

$$p(\beta_l \mid \alpha_k)p(\alpha_k) = p(\alpha_k \mid \beta_l)p(\beta_l) \tag{1.126}$$

From (1.124) and (1.126) it follows that

$$p(\alpha_k \mid \beta_l) = p(\alpha_k) \tag{1.127}$$

The events α_k and β_l satisfying the equivalent conditions expressed by (1.124) or (1.125) or (1.127) are called statistically independent events. To use some statistical terminology, we see that if the joint density $p(\alpha_k, \beta_l)$ can be factored into the product of the two marginal densities $p(\alpha_k)$ and $p(\beta_l)$, the events α_k and β_l are statistically independent. Here, we recall that the marginal densities are constructed from the joint density by a summation of the type shown in (1.118).

1.26 Random Variable

First let us define the sample point s as a point associated with each possible outcome of an experiment. A random variable is a real-valued function $x(s)$ defined for each s for which a probability density function is defined. For example, in the experiment of throwing a die the outcomes are the sides identified by 1, 2, . . . , 6. These are the sample points. The numbers 1, 2, . . . , 6 can be chosen as random variables (number 1 chosen if side 1 appears, etc.). Hence, $x(1) = 1$, $x(2) = 2, . . .$, etc. The function $x(k) = k$ is a random variable. The functions $y(k) = k^3$, $z(k) = \exp(k)$ are also random variables. A random variable can be scalar-valued, as in the example above, or vector-valued such as a vector representing the coordinates of a randomly moving particle in space.

A random variable may be discrete or continuous. In the example of throwing a die, the random variable can take on any of the six possible values while any component of the vector representing the position of a random particle at any time t might assume any value between plus and minus infinity. In the former case we have a discrete and in the latter a continuous random variable.

The definitions and results of Sections 1.24 and 1.25 are directly applicable to random variables. For example, the discrete random variables x and y having K and L values $x_1, \ldots, x_K$, and $y_1, \ldots, y_L$ respectively are called statistically independent if

$$p(x_k, y_l) = p(x_k)p(y_l) \qquad \text{for all } k \text{ and } l \qquad (1.128)$$

where $p(x_k, y_l)$ is the joint probability density function of x_k and y_l. Similarly, if x and y are continuous random variables, they are called statistically independent if

$$p(x, y) = p(x)p(y) \qquad \text{for all } x \text{ and } y \qquad (1.129)$$

1.27 Moments

Let x be a scalar random variable. The nth moment of x is defined by

$$m_n \triangleq Ex^n \triangleq \int_{-\infty}^{+\infty} x^n p(x)\, dx \qquad (1.130)$$

The nth central moment of x is defined by

$$\mu_n \triangleq E[x - Ex]^n \triangleq \int_{-\infty}^{+\infty} [x - Ex]^n p(x)\, dx$$

The first moment of x

$$m_1 \triangleq Ex \triangleq \int_{-\infty}^{+\infty} x p(x)\, dx$$

is also called the mean of x. In general, by expected value of a function of several arguments, such as $f(x, y, z)$, is meant

$$Ef(x, y, z) = \int\!\!\int\!\!\int_{-\infty}^{+\infty} f(x, y, z)p(x, y, z)\, dx\, dy\, dz$$

The definition of a moment applies to discrete random variables if we simply substitute summations in place of integrations in the definitions. The expected value may be a scalar or a vector, depending on whether the function $f(x, y, z)$ is scalar- or vector-valued. The integration is understood to be applied to each element of the vector function $f(x, y, z)$ when f is vector-valued.

▶ example

Let the probability density function of x be uniform over the interval $[0, 1]$

$$p(x) = 1 \qquad 0 \leq x \leq 1$$
$$= 0 \qquad x < 0, \ x > 1$$

Then the various moments of x become

$$m_n = Ex^n = \int_0^1 x^n \, dx = \frac{1}{n+1}$$

$$m_1 = Ex = \tfrac{1}{2}$$

and, since $m_1 = \tfrac{1}{2}$, the nth central moment is given by

$$\mu_n = E[x - \tfrac{1}{2}]^n = \int_0^1 [x - \tfrac{1}{2}]^n \, dx$$

Hence

$$\mu_n = \frac{1}{(n+1)2^n} \qquad n = 2, 4, 6, 8, \ldots$$
▶
$$= 0 \qquad n = 1, 3, 5, 7, \ldots$$

It is important to realize that some of the moments associated with certain probability density functions $p(x)$ can fail to exist, because the defining integral does not exist. This pathological situation only occurs for *some* probability density functions which are infinite at certain values of x or which are defined for an infinite range of x.

1.28   Random Processes

A random variable was defined as a function $x(s)$ defined for each outcome of an experiment identified by s. Now if we assign to each outcome s a time function $x(t, s)$ we obtain a family of functions called a "stochastic process" or "random process." Normally, the explicit dependence on s is not shown, and a random process is given as a function of one argument t, such as $x(t)$. From this definition it is clear that the value of a random process $x(t)$ at any particular time $t = t_0$, namely $x(t_0)$, is a random variable (or random vector if $x(t)$ is vector-valued). A random process is called discrete if its argument is a discrete variable such as

$$x(k), \qquad k = 1, 2, \ldots$$

An example of a continuous random process is the noise voltage as a function of time at the output of a radio receiver. The noise voltage at a particular time is a random variable. If we sample the noise voltage, say each second, then we have generated a discrete random process (or a random sequence).

The definitions and results of Sections 1.24 and 1.25 are applicable to random processes. For example, if $x(k), k = 1, \ldots, K$ and $y(l), l = 1, \ldots, L$ are two discrete processes, then they are called statistically independent if

$$p[x(1), \ldots, x(K), y(1), \ldots, y(L)] = p[x(1), \ldots, x(K)]p[y(1), \ldots, y(L)]$$

The Bayes' Formula given by (1.120) takes the form

$$p[x(1), \ldots, x(K) \mid y(1), \ldots, y(L)] = \frac{p[x(1), \ldots, x(K), y(1), \ldots, y(L)]}{p[y(1), \ldots, yL)]}$$

$$(1.131)$$

where the left hand side is the conditional density function of $x(k), k = 1, \ldots, K$, given $y(l)$ $l = 1, \ldots, L$.

When $x(t)$ is vector-valued (say an n-vector), then by the probability density $p[x(t)]$ is meant $p[x_1(t), x_2(t), \ldots, x_n(t)]$ and by the joint density $p[x(t_1), y(t_2)]$ is meant $p[x_1(t_1), \ldots, x_n(t_1), y_1(t_2), \ldots, y_m(t_2)]$ where $y(t)$ is an m-vector.

1.29 The Statistics of Processes

We can consider the discrete random process $x(k), k = 1, \ldots, K$ by a set of random variables $x(1), x(2), \ldots, x(K)$. The simplest statistical information among these variables is the probability density function of each individual variable. This information is called the "first order statistics" of $x(k)$ and is specified by

$$p[x(k)], \qquad k = 1, \ldots, K$$

The "second order statistics" is specified by the following joint probability density function involving values of the random process at all possible *pairs* of indices k_1 and k_2

$$p[x(k_1), x(k_2)] \qquad k_1 = 1, \ldots, K; \qquad k_2 = 1, \ldots, K$$

In the same manner the "nth order statistics" is specified by the joint probability density function involving values of the random process at any set of n indices

$$p[x(k_1), x(k_2), \ldots, x(k_n)] \qquad k_1, k_2, \ldots, k_n = 1, 2, \ldots, K$$

Similarly, the first, second, and nth order statistics of a continuous random process are defined by the following probability functions respectively:

$$p[x(t)]$$
$$p[x(t_1), x(t_2)]$$
$$p[x(t_1), x(t_2), \ldots, x(t_n)]$$

A stochastic process $x(t)$ is statistically determined if its nth order probability density functions are known for all n, and for all assignments of time points $t_1, t_2, \ldots, t_n$. Note that the density function of order k can be obtained from the density function of order n if $n > k$ by integrating (or summing) over the variables which are not needed. For example,

$$p[x(t_1), x(t_2)] = \int_{-\infty}^{+\infty} p[x(t_1), x(t_2), x(t_3)] \, dx(t_3) \qquad (1.132)$$

The integration in (1.132) obviously obtains the marginal density on $x(t_1)$ and $x(t_2)$ from the joint density on $x(t_1)$, $x(t_2)$, $x(t_3)$, as was done in discrete form in (1.118).

1.30   Mean, Correlation, Covariance

Let $x(t)$ (or $x(k)$) be an n-vector random process (i.e., a vector consisting of n components, each a random process). The mean of the random process $x(t)$ is denoted by $Ex(t)$ and is defined as follows

$$Ex(t) \triangleq \int_{-\infty}^{+\infty} x(t)p[x(t)] \, dx(t) \qquad (1.133)$$

which is to be interpreted elementwise as

$$Ex_i(t) \triangleq \int_{-\infty}^{+\infty} x_i(t)p[x_i(t)] \, dx_i(t) \qquad i = 1, \ldots, n$$

In (1.133), it is understood that the value of t at which integration is carried out is held fixed. Of course, the value of $Ex(t)$ is, in general, a function of t.

The autocorrelation of the vector-valued process $x(t)$ is defined by the following matrix, which has the form of an outer product:

$$Ex(t_1)x'(t_2) \triangleq \begin{bmatrix} Ex_1(t_1)x_1(t_2) & Ex_1(t_1)x_2(t_2) & \cdots \\ Ex_2(t_1)x_1(t_2) & Ex_2(t_1)x_2(t_2) & \\ \cdot & \cdot & \\ \cdot & & \cdot \\ \cdot & & & \cdot \end{bmatrix} \qquad (1.134)$$

where

$$Ex_i(t_1)x_j(t_2) = \int\!\!\int_{-\infty}^{+\infty} x_i(t_1)x_j(t_2)p[x_i(t_1), x_j(t_2)] \, dx_i(t_1) \, dx_j(t_2) \qquad (1.135)$$

The covariance of $x(t)$ is defined by

$$E[x(t_1) - Ex(t_1)][x(t_2) - Ex(t_2)]' = E[x(t_1)x'(t_2)] - Ex(t_1)Ex'(t_2) \qquad (1.136)$$

It may be seen that when $x(t)$ has zero mean, $Ex(t_1) = 0$, its auto-correlation and covariance are identical.

The correlation matrix of two random processes $x(t)$, an n-vector, and $y(t)$, an m-vector, are given by the $n \times m$ matrix

$$Ex(t_1)y'(t_2)$$

where

$$Ex_i(t_1)y_j(t_2) = \int\int_{-\infty}^{+\infty} x_i(t_1)y_j(t_2)p[x_i(t_1), y_j(t_2)] \, dx_i(t_1) \, dy_j(t_2) \quad (1.137)$$

Similarly, the covariance matrix is an $n \times m$ matrix given by

$$E[x(t_1) - Ex(t_1)][y(t_2) - Ey(t_2)]' \quad (1.138)$$

▶ **example**

Let α and β be two n-dimensional random vectors where $E\alpha$, $E\beta$, $E\alpha\beta'$ are given. The function

$$x(t) = \alpha + \beta t$$

then represents a random process. The mean value of $x(t)$ follows from (1.133) as

$$Ex(t) = E\alpha + [E\beta]t$$

The autocorrelation matrix becomes

$$Ex(t_1)x'(t_2) = E\alpha\alpha' + [E\alpha\beta']t_2 + [E\beta\alpha']t_1 + [E\beta\beta']t_1t_2 \quad (1.139)$$

The covariance matrix of $x(t)$ is given by

$$\begin{aligned}
E[x(t_1) &- Ex(t_1)][x(t_2) - Ex(t_2)'] \\
&= E\alpha\alpha' + [E\alpha\beta']t_2 + [E\beta\alpha']t_1 + [E\beta\beta']t_1t_2 \\
&- E\alpha E\alpha' - [E\beta E\beta']t_1t_2 - [E\alpha E\beta']t_2 - [E\beta E\alpha']t_1 \quad (1.140]
\end{aligned}$$

Clearly if $E\alpha = E\beta = 0$ the right hand side of (1.139) and (1.140) become ▶ identical.

1.31 Orthogonal Processes

Two random processes $x(t)$ and $y(t)$ are called uncorrelated if their co-variance matrix is identically zero, for all t_1 and t_2

$$E[x(t_1) - Ex(t_1)][y(t_2) - Ey(t_2)]' = 0 \quad (1.141)$$

The processes $x(t)$ and $y(t)$ are called orthogonal if in addition they have zero mean; in other words, if their correlation matrix is identically zero:

$$Ex(t_1)y'(t_2) = 0 \quad (1.142)$$

The random process $x(t)$ is called uncorrelated if

$$E[x(t_1) - Ex(t_1)][x(t_2) - Ex(t_2)]' = K(t_1, t_2)\delta(t_1 - t_2)$$

where $\delta(t)$ is the Dirac delta function, namely

$$\delta(t) = 0 \qquad t \neq 0$$

$$\int_{-\infty}^{+\infty} \delta(t)\, dt = 1$$

Similarly, the random sequence $x(k)$ is called uncorrelated if

$$E[x(k_1) - Ex(k_1)][x(k_2) - Ex(k_2)]' = K(k_1, k_2)\, \Delta(k_2 - k_1)$$

where $\Delta(k_2 - k_1)$ is the Kronecker delta function, namely,

$$\Delta(k_2 - k_1) = 0 \qquad k_1 \neq k_2$$
$$\Delta(k_2 - k_1) = 1 \qquad k_1 = k_2$$

The so-called "white noise" studied in communications and estimation theory is an example of an uncorrelated process. In the above definition of an uncorrelated process, we see that it is the *covariance* (not correlation) which is specified in terms of a delta function.

1.32 Stationary Processes

The random process $x(t)$ $\big($or random sequence $x(k)\big)$ is called stationary (in the strict sense) if *all* its statistics $\big($meaning $p[x(t_1), x(t_2), \ldots]\big)$ are invariant with respect to a shift in time origin. In other words, if all the statistics of the random processes $x(t)$ and $x(t + \tau)$ are identical for any τ, $-\infty \leq \tau \leq +\infty$, then the process or sequence is strictly stationary. In terms of random sequences this means that the statistics of $x(k)$ and $x(k + m)$ are identical for all allowable values of m (for example, when k is an integer, m is then any integer).

The random process $x(t)$ $\big($or $x(k)\big)$ is called stationary in the wide sense if its expected value is constant and its autocorrelation is only a function of the difference $t_2 - t_1$

$$Ex(t) = \text{Constant} \tag{1.143}$$

$$Ex(t_1)x'(t_2) = K(t_2 - t_1) \tag{1.144}$$

where K is a matrix with each element depending only on the difference $t_2 - t_1$ rather than on some more complicated relationship involving t_1 and t_2.

Let us denote the functions representing $p[x(t)]$ and $p[x(t_1), x(t_2)]$ by $f(x, t)$ and $f(x_1, x_2, t_1, t_2)$ respectively. Then from (1.143), if $x(t)$ is stationary, we must necessarily have

$$Ex(t) = \int_{-\infty}^{+\infty} xf(x, t) \, dx = \text{Constant (independent of } t) \qquad (1.145)$$

Hence $f(x, t)$ must be independent of time

$$f(x, t) = f(x) \qquad (1.146)$$

Furthermore, from (1.144), where for simplicity in notation $x(t)$ is assumed scalar, we have

$$Ex(t_1)x(t_2) = \int_{-\infty}^{+\infty}\int x_1 x_2 f(x_1, x_2, t_1, t_2) \, dx_1 \, dx_2 = K(t_2 - t_1)$$

Since the only term in the integrand which is a function of t_1 and t_2 is $f(x_1, x_2, t_1, t_2)$ it must necessarily have the following form

$$f(x_1, x_2, t_1, t_2) = f(x_1, x_2, t_2 - t_1)$$

Therefore, when $x(t)$ is stationary in the wide sense, it implies that its first and second order statistics are independent of time origin, while strict stationarity by definition implies that statistics of all orders are independent of time origin. Wide sense stationary processes are often considered in engineering applications.

▶ example
Let a and b be scalar random variables and λ be a constant. Consider the process $x(t)$ given by

$$x(t) = a \cos \lambda t + b \sin \lambda t$$

which is a random process by definition. We have, for each t,

$$Ex(t) = [Ea] \cos \lambda t + [Eb] \sin \lambda t$$

Furthermore, for each t_1 and t_2,

$$Ex(t_1)x(t_2) = \tfrac{1}{2}[Ea^2][\cos \lambda(t_2 + t_1) + \cos \lambda(t_2 - t_1)]$$
$$+ \tfrac{1}{2}[Eb^2][-\cos \lambda(t_2 + t_1) + \cos \lambda(t_2 - t_1)]$$
$$+ [Eab] \sin \lambda(t_1 + t_2)$$

Consequently, in general, $x(t)$ is a nonstationary process. However, if the random variables a and b were to have the properties

$$Ea = Eb = Eab = 0; \qquad Ea^2 = Eb^2$$

then we would obtain

$$Ex(t) = 0$$
$$Ex(t_1)x(t_2) = [Ea^2] \cos \lambda(t_2 - t_1)$$

▶ which certainly represents a stationary process in the wide sense.

1.33 Ergodic Processes

The expected values used in the preceding section and denoted by the symbol E are called ensemble averages. The reason for this is that the operation implies that the averaging is over the response of many similar (or statistically identical) experiments evaluated at the same time (or times). For example, by $Ex(t_1)$ we mean that for a set of statistically identical experiments producing sample outputs $x(t)$, each response is sampled at the same time $t = t_1$ and then the results averaged.

A completely different type of average can be defined by only one response (outcome of a single experiment). Such an average is called a time average for which we use the symbol E_t. The first two time averages of the random process $x(t)$ are defined by

$$E_t[x(t)] = \frac{1}{2T} \int_{-T}^{+T} x(t) \, dt \qquad (1.147)$$

and

$$E_t[x(t)x(t + \tau)] = \frac{1}{2T} \int_{-T}^{+T} x(t)x(t + \tau) \, dt \qquad (1.148)$$

where T is a constant. Clearly, $E_t[x(t)]$ and $E_t[x(t)x(t + \tau)]$ are random variables by themselves owing to their dependence on $x(t)$. Consequently, we can obtain their ensemble expected values:

$$EE_t[x(t)] = \frac{1}{2T} \int_{-T}^{T} Ex(t) \, dt$$

and

$$EE_t[x(t)x(t + \tau)] = \frac{1}{2T} \int_{-T}^{T} Ex(t)x(t + \tau) \, dt$$

Now if $x(t)$ is a stationary process, i.e., $Ex(t)$ and $Ex(t)x(t + \tau)$ are independent of time origin, it follows that

$$EE_t[x(t)] = Ex(t) \qquad (1.149)$$

and

$$EE_t[x(t)x(t + \tau)] = Ex(t)x(t + \tau) \qquad (1.150)$$

If the limits of the time averages given by (1.147) and (1.148) as $T \to \infty$ exist and are constants independent of any chosen sample function $x(t)$, then (1.149) and (1.150) indicate that the limit of time averages and ensemble averages are identical, and the process is called "ergodic in the mean and autocorrelation." It can be shown that if

$$\lim_{T \to \infty} \frac{1}{2T} \int_{-T}^{T} x(t)\, dt$$

exists,* then the process is ergodic in the mean.

A physical interpretation of ergodicity is that one sample of the random process $x(t)$ as a function of time eventually takes on values (as t approaches infinity) with the same statistics as the value of say $x(t_1)$ corresponding to many outcomes at the same time, of statistically identical experiments. Ergodicity is an important property of a random process, since it allows us to infer statistical information concerning the ensemble of process outputs $x(t)$ from statistical information based on a single observation made over a long interval of time. For example, if $x(t)$ is a scalar ergodic process, then the statistical average $Ex^2(t)$ is equal to the limit of the time average of $x^2(t)$:

$$Ex^2(t) = \lim_{T \to \infty} \frac{1}{2T} \int_{-T}^{+T} x^2(t)\, dt$$

where the left hand side is usually defined or referred to as the average energy of $x(t)$. From (1.144), or in general, when the process $x(t)$ is stationary or at least wide-sense stationary, the left hand side is a constant (independent of time t).

1.34 Functions of Random Variables and Processes

A function of a random variable x is the operation of assigning to each value of x another value, say y, according to some rule or function. This is denoted in general by

$$y = f(x) \tag{1.151}$$

where x and y are usually called input and output respectively. If the value of y depends on both x and time t, the function is called time-varying

$$y = f(x, t) \tag{1.152}$$

It is of great importance in practice to determine the statistical properties of y in terms of those of x. For example, we could consider the various

---

* This can always be done for stationary processes. *See* Middleton, *Statistical Communication Theory.*

moments of y. By definition we have the following expressions for the moments of y

$$Ey = \int_{-\infty}^{+\infty} f(x)p(x)\,dx$$

$$Ey^2 = \int_{-\infty}^{+\infty} [f(x)]^2 p(x)\,dx$$

$$
\begin{array}{cc}
\cdot & \cdot \\
\cdot & \cdot \\
\cdot & \cdot
\end{array}
$$

when y is a scalar. For vector-valued functions y, similar expressions can be identified. It may be seen that the nth moment of y may involve any number of moments of x. For example, if x is a scalar and

$$y = x^2$$

then

$$Ey = Ex^2$$

As another example, if

$$y = \frac{x}{1-x}$$

where x is uniformly distributed over the interval 0, $\frac{1}{2}$ then

$$Ey = Ex + Ex^2 + Ex^3 + \cdots$$

$$= \sum_{n=1}^{\infty} Ex^n$$

In general, the probability density of y can be obtained from the density of x. Specifically when (1.151) can be solved for x yielding the unique solution

$$x = g(y)$$

then we have

$$p_y(y) = \{p_x[x = g(y)]\} \frac{1}{\left| \dfrac{\partial f(x)}{\partial x} \right|_{x=g(y)}} \tag{1.153}$$

where $p_y(y)$ and $p_x(x)$ are the density functions of y and x respectively. An outline of the proof is given in the following.* The probability that y lies between y_1 and $y_1 + dy$ is $p_y(y)\,dy$. From (1.151) and $x = g(y)$ this corresponds to x occurring between x_1 and $x_1 + dx$ which has the probability

* For details see *Papoulis, Probability, Random Variables, and Stochastic Processes.*

$p_x(x)\,dx$. Furthermore, from (1.151) we have $dy = \dfrac{\partial f}{\partial x}\,dx$. Hence

$$p_y(y)\,dy = p_x(x)\,\frac{1}{\left|\dfrac{\partial f}{\partial x}\right|}\,dy\,\Bigg|_{x=g(y)}$$

A function of two random variables x, y is the process of assigning to each pair of x, y another value, say z according to some rule.

$$z = f(x, y)$$

Given the probability density function of $p(x, y)$ we often need to derive the density of z. In general, this is rather complicated. In the special case where z is linear in x and y and where x, y are independent random variables a simple relationship for $p(z)$ can be derived.

Let $p(x) = f_x(x)$, $p(y) = f_y(y)$, and $p(x, y) = p(x)p(y)$. Furthermore,

$$z = \alpha x + \beta y$$

where α and β are constants. The probability that z assumes a value such as z_0 is the collection of the joint events where

$$\beta x = \xi$$
$$\beta y = z_0 - \xi$$

for all possible values of ξ. Since x and y are independent, the probability of the joint event is

$$p(\alpha x = \xi, \beta y = z_0 - \xi) = \frac{1}{|\alpha\beta|}\,f_x(\xi/\alpha)f_y\!\left(\frac{z_0 - \xi}{\beta}\right)$$

Since these events are mutually exclusive, we have

$$p(z) = \frac{1}{|\alpha\beta|}\int_{-\infty}^{+\infty} f_x(\xi/\alpha)f_y\!\left(\frac{z_0 - \xi}{\beta}\right)d\xi$$

For example, if $\alpha = \beta = 1$, this indicates that the probability of the sum of two independent random variables is the convolution of their probability density functions.

When x and y in (1.151) are n-dimensional vectors and if a unique solution for x in terms of y exists, namely

$$x = g(y)$$

then (1.153) becomes

$$p_y(y) = \{p_x[x = g(y)]\}\,\frac{1}{|J|_{x=g(y)}}$$

where $|J|$ is the Jacobian defined by the

$$|J| = \begin{vmatrix} \dfrac{\partial f_1}{\partial x_1} & \dfrac{\partial f_2}{\partial x_1} & \cdots & \dfrac{\partial f_n}{\partial x_1} \\[2ex] \cdot & & & \cdot \\ \cdot & & & \cdot \\ \cdot & & & \cdot \\[1ex] \dfrac{\partial f_1}{\partial x_n} & & & \dfrac{\partial f_n}{\partial x_n} \end{vmatrix}$$

A function of a random process is defined in the same manner as was done for a function of a random variable

$$y(t) = f[x(t), t] \tag{1.154}$$

and the function $y(t)$ becomes, of course, another random process. By definition of a random process, the quantities $x(t_0)$ and $y(t_0)$ are random variables for any particular value of time such as t_0. However, in the case of random processes, there is a more general form for the output $y(t)$, namely, when $y(t)$ is related not only to $x(t)$ at t but to the value of the random process $x(t)$ at a number of different times $t_1, t_2, \ldots$, or even over an interval of time. Such a functional relationship between $x(t)$ and $y(t)$ is referred to as "having memory," in contrast with the relationship (1.154) which is called a "memoryless function."

A very common situation in practice which leads to functions with memory is when the processes $x(t)$ and $y(t)$ (or $x(k)$ and $y(k)$) are related to each other through a differential (or difference) equation as input and output respectively. These are the type of processes with which we will be involved throughout this book. The special case where such differential (or difference) equations are linear is discussed in the next section.

1.35 Linear Continuous Systems

Let $u(t)$ be an r-dimensional vector random process with the following statistical properties:

$$Eu(t) = m(t) \tag{1.155}$$

$$E[u(t_1) - m(t_1)][u(t_2) - m(t_2)]' = K(t_2, t_1) \tag{1.156}$$

where $K(t_2, t_1)$ is an $r \times r$ matrix which is the covariance matrix of $u(t)$. Now let the n-vector $x(t)$ be given as the solution of the following vector linear differential equation.

$$\dot{x}(t) = A(t)x(t) + B(t)u(t)$$
$$x(0) = x_0 \tag{1.157}$$

We assume that to each $u(t)$ and for a given x_0 there corresponds a unique function $x(t)$ satisfying (1.157). The solution to (1.157) may be written in the form (1.92):

$$x(t) = X(t)x_0 + \int_0^t X(t)X^{-1}(s)B(s)u(s)\,ds; \qquad t \geq 0 \qquad (1.158)$$

where $X(t)$ is the transition matrix and s is the dummy variable of integration. Notice that the only random term in the right hand side of (1.158) is the random process $u(s)$. Consequently, by taking the expectation of both sides and interchanging the order of integration with respect to s and the operation of expectation, it follows that

$$Ex(t) = X(t)x_0 + \int_0^t X(t)X^{-1}(s)B(s)Eu(s)\,ds \qquad (1.159)$$

This implies, for a linear differential equation such as (1.157), that in order to obtain the mean of the random process $x(t)$, we can drive the differential equation by the mean of the input random process $u(t)$ $[Eu(t) = m(t)]$.

Next let us derive an expression for the covariance of the random process $x(t)$ in terms of the covariance $K(t_2, t_1)$ of the input random process $u(t)$. We have, by definition (1.138)

$$\text{Cov}\,[x(t_1), x(t_2)] \triangleq E[x(t_1) - Ex(t_1)][x(t_2) - Ex(t_2)]' \qquad (1.160)$$

From (1.158) and (1.159) we have

$$x(t) - Ex(t) = \int_0^t X(t)X^{-1}(s)B(s)[u(s) - Eu(s)]\,ds \qquad (1.161)$$

Substituting (1.161) into (1.160), transposing, then writing the iterated integral as a double integral and interchanging the integration and expectation operations yields

$$\text{Cov}\,[x(t_0), x(t_2)] = \int_0^{t_1} \int_0^{t_2} X(t_1)X^{-1}(s)B(s)E\{[u(s) - Eu(s)]$$
$$\times\,[u(\sigma) - Eu(\sigma)]'B'(\sigma)X^{-1}(\sigma)X'(t_2)\,d\sigma\,ds \qquad (1.162)$$

which becomes, on utilizing (1.156)

$$\text{Cov}\,[x(t_1), x(t_2)] = \int_0^{t_1} \int_0^{t_2} X(t_1)X^{-1}(s)B(s)K(\sigma, s)B'(\sigma)X^{-1'}(\sigma)X'(t_2)\,d\sigma\,ds$$
$$(1.163)$$

This is the desired result in general form. Note that the covariance (1.163) depends on t_1 and t_2, but not on the initial state x_0. Now let the covariance matrix $K(t_1, t_2)$ have the very special form, namely,

$$K(s, \sigma) = Q(s, \sigma)\delta(\sigma - s) \qquad (1.164)$$

where the dummy variables s and σ have been substituted* for t_1 and t_2 in the definition (1.156), and where $\delta(t)$ is the Dirac delta function, i.e.,

$$\sigma(t) = 0 \qquad t \neq 0$$

$$\int_{-\infty}^{+\infty} \delta(t)\, dt = 1 \qquad (1.165)$$

The Dirac delta function has the "sifting property" that for any function $z(t)$

$$\int_{-\infty}^{+\infty} z(\tau)\, \delta(\tau)\, d\tau = z(0)$$

Substituting (1.164) into (1.163), we can hold s fixed and perform the integration with respect to σ. By the sifting property of the delta function, and by associating the dummy variable σ with the integration from 0 to t_2 where $t_2 > t_1$, (1.163) will simplify considerably as follows

$$\text{Cov}\,[x(t_1),\, x(t_2)] = \int_0^{t_1} X(t_1)X^{-1}(s)B(s)Q(s,s)B'(s)X^{-1'}(s)X'(t_2)\, ds \quad (1.166)$$

As will become apparent in the succeeding chapters, we shall attempt to represent our mathematical models of random processes so that the known processes appear as uncorrelated processes in the form (1.164). We remark that independent, zero-mean processes† satisfy the desired condition (1.164).

Let us return to (1.157) and consider another special case, namely, when A is a constant matrix, $K(t_1, t_2)$ is given as $K(t_2 - t_1) = \varphi_u(t_2 - t_1)$, i.e., $u(t)$ is stationary (with covariance function φ_u) and B is an $n \times 1$ constant vector. Furthermore, let

$$y(t) = Cx(t) \qquad (1.167)$$

where C is a $1 \times n$ vector. According to (1.102) and (1.105), $u(t)$ and $y(t)$ are related by the convolution integral

$$y(t) = \int_0^t h(t - s)u(s)\, ds \qquad (1.168)$$

where

$$h(t) = Ce^{At}B \qquad (1.169)$$

is the impulse response.

---

* The arguments t_1 and t_2 of the function $K(t_1, t_2)$ are not to be confused with the limits of integration, which also use the notation t_1 and t_2.

† The process $x(t)$ is independent if

$$p[x(t_1), x(t_2)] = p[x(t_1)]p[x(t_2)] \qquad \text{for any} \qquad t_1 \neq t_2$$

cf., expression (1.129). Hence when $Ex(t_1) = Ex(t_2) = 0$, then

$$\text{Cov}\,[x(t_1), x(t_2)] = E[x(t_1)]E[x'(t_2)] = 0 \qquad \text{for} \qquad t_2 \neq t_1$$

We may now derive the covariance of $y(t)$. Note that the usage of the covariance $K(t_2 - t_1) = \varphi_u(t_2 - t_1)$ implies that all effects due to the mean, $Eu(t)$, have been subtracted out, as in (1.161). From (1.167) it follows that

$$Ey(t)y(t + \tau) = CEx(t)x'(t + \tau)C' \tag{1.170}$$

Substituting for $Ex(t)x'(t + \tau)$ from (1.163) by letting $t_1 = t$, $t_2 = t + \tau$, $X(t) = e^{At}$ it follows that

$$Ey(t)y(t + \tau) = \int_0^t \int_0^{t+\tau} h(t - s)\varphi_u(\sigma - s)h(t + \tau - \sigma)\,d\sigma\,ds \tag{1.171}$$

Using the change of variables, for t and τ held constant,

$$t - s = \alpha$$

and

$$t + \tau - \sigma = \beta$$

Equation (1.171) becomes

$$Ey(t)y(t + \tau) = \int_0^\alpha \int_0^{t+\tau} h(\alpha)h(\beta)\varphi_u(\tau + \alpha - \beta)\,d\beta\,d\alpha$$

For the special case where the input $u(t)$ in (1.157) is specified over the $[-\infty, t]$ interval (and if $x(0) = 0$ in (1.100) is now replaced by the condition that $h(t)$ be stable), and when $h(t) = 0$, $t \leq 0$ (i.e., when $h(t)$ is, by definition, "causal") (1.171) becomes

$$\lim_{t\to\infty} Ey(t)y(t + \tau) \triangleq \varphi_{yy}(\tau) = \int_{-\infty}^{+\infty} \int_{-\infty}^{+\infty} h(\alpha)h(\beta)\varphi_u(\tau + \alpha - \beta)\,d\beta\,d\alpha \tag{1.172}$$

Equation (1.172) is meaningful only if the limit indicated in the left hand side exists.

A sufficient condition for the existence of $\varphi_{yy}(\tau)$ can easily be derived. By simple expansion, we have (assuming $y(t)$ is stationary)

$$E[y(t) + \alpha y(t + \tau)]^2 = Ey^2(t) + 2\alpha Ey(t)y(t + \tau) + \alpha^2 Ey^2(t)$$

where α is an arbitrary constant. Since the left hand side is non-negative for any α it follows that

$$Ey^2(t) + 2\alpha Ey(t)y(t + \tau) + \alpha^2 Ey^2(t) \geq 0$$

or by definition of autocorrelation $\varphi_{yy}(\tau)$ (equation (1.172))

$$\varphi_{yy}(0) + 2\alpha\varphi_{yy}(\tau) + \alpha^2\varphi_{yy}(0) \geq 0$$

Since this must necessarily be satisfied for all α, it is necessary that

$$\varphi_{yy}^2(\tau) \leq \varphi^2(0)$$

or

$$|\varphi_{yy}(\tau)| \leq Ey^2(t) \tag{1.173}$$

The right hand side of (1.173) may be identified as the average energy of the process $y(t)$. Thus, φ_{yy} exists and is bounded when the average energy of $y(t)$ is finite. For example, with the ergodicity condition we have

$$Ey^2(t) = \lim_{T \to \infty} \frac{1}{2T} \int_{-T}^{+T} y^2(t)\, dt$$

Consequently, the value of $|\varphi_{yy}(\tau)|$ is bounded if the average energy in the process $y(t)$ is finite.

It is interesting to notice that, as a direct consequence of the definition of $\varphi(\tau)$, the autocorrelation function (and also the covariance function) for stationary processes is an even function, i.e.,

$$\varphi(\tau) = \varphi(-\tau)$$

▶ example 1

A scalar random process $x(t)$ is given by the solution of the following differential equation

$$\dot{x}(t) = -(1 + t)x(t) + u(t)$$
$$x(0) = 1 \qquad\qquad (1.174)$$

where

$$Eu(t) = 1$$
$$E[u(t_1) - 1][u(t_2) - 1] = \delta(t_2 - t_1)$$

The transition matrix (here a scalar quantity) corresponding to (1.174) is easily derived as

$$X(t) = e^{-[t+(1/2)t^2]}$$

The expected value of $x(t)$ from (1.159) is given by

$$Ex(t) = e^{-[t+(1/2)t^2]}\left[1 + \int_0^t e^{[s+(1/2)s^2]}\, ds\right]$$

and its covariance is obtained from (1.166) as

$$\text{Cov } [x(t_1)x(t_2)] = e^{-[t_1+(1/2)t_1^2+t_2+(1/2)t_2^2]} \int_0^{t_1} e^{2(s+(1/2)s^2)}\, ds \qquad t_2 \geq t_1$$

Notice that

$$\lim_{t_2 \to \infty} \text{Cov } [x(t_1), x(t_2)] \to 0$$

which indicates that the correlation between any two time samples ▶ approaches zero as the time interval tends to infinity.

▶ **example 2**

Let a scalar differential equation be given by

$$\dot{x}(t) = x(t) + u(t)$$
$$x(0) = 0 \qquad (1.175)$$

where

$$Eu(t) = 0$$
$$Eu(t_1)u(t_2) = \delta(t_2 - t_1)$$

Clearly, from (1.159),

$$Ex(t) = 0$$

and for $\tau \geq 0$, from (1.166), we can identify $t = t_1$ and $t + \tau = t_2$ to obtain

$$E[x(t)x(t + \tau)] = \int_0^t e^{2t}e^{\tau}e^{-2s}\,ds = \tfrac{1}{2}e^{\tau}[e^{2t} - 1]$$

and since a similar value can be obtained for $\tau < 0$ as long as $t, t + \tau \geq 0$, then

$$E[x(t)x(t + \tau)] = \tfrac{1}{2}e^{|\tau|}[e^{2t} - 1]$$

which indicates that the covariance function is zero at $t = 0$ and approaches infinity with t. Both results are intuitively evident: the former since the system started with zero initial conditions; the latter owing to the fact that the system given by (1.175) represents an unstable differential equation. Had we chosen

$$\dot{x}(t) = -x(t) + u(t)$$

in place of (1.175), the resultant covariance function would have been

$$Ex(t)x(t + \tau) = \tfrac{1}{2}e^{-|\tau|}[1 - e^{-2t}]$$

which again yields a value of zero at $t = 0$, but its limit as t approaches infinity exists

$$\lim_{t \to \infty} Ex(t)x(t + \tau) = \tfrac{1}{2}e^{-|\tau|} \qquad (1.176)$$

This implies that the process $x(t)$ is asymptotically (i.e., as $t \to \infty$) wide sense stationary, because at $t = \infty$ the dependence of $Ex(t)x(t + \tau)$ on t disappears. This limiting value can also be obtained from (1.172). From (1.175) we have

$$\varphi_u(\tau) = \delta(\tau)$$

and using, $\dot{x}(t) = -x(t) + u(t)$ we have

$$h(t) = e^{-t} \quad t \geq 0$$
$$= 0 \qquad t < 0$$

Hence

▶ $$\varphi_{vv}(\tau) = \int_0^{+\infty} \int_0^{+\infty} e^{-\alpha}e^{-\beta}\, \delta(\tau + \alpha - \beta)\, d\beta\, d\alpha = \int_0^{+\infty} e^{-\alpha}e^{-\alpha-\tau}\, d\alpha$$
$$= \tfrac{1}{2}e^{-|\tau|}$$

1.36 Linear Discrete Systems

Let the input $u(t)$ to the difference equation

$$x(k+1) = A(k)x(k) + B(k)u(k)$$
$$x(0) = x_0 \tag{1.177}$$

be an r-dimensional random vector where

$$Eu(k) = m(k) \tag{1.178}$$

$$E[u(k_1) - m(k_1)][u(k_2) - m(k_2)]' = K(k_2, k_1) \tag{1.179}$$

In order to determine the mean and covariance of the solution $x(k)$, namely, $Ex(k)$ and $E[x(k_1) - Ex(k_1)][x(k_2) - Ex(k_2)]'$, we can use a procedure similar to that of the preceding section by using the solution to linear difference equation (1.177) (*see* Section 1.22).

A different approach which often has some computational advantages is to derive deterministic equations whose solutions are the desired mean and and covariance matrix of $x(k)$. Taking the expectation of both sides of (1.177) yields

$$Ex(k+1) = A(k)Ex(k) + B(k)Eu(k)$$
$$Ex(0) = x_0 \tag{1.180}$$

This is a deterministic linear difference equation which yields as its solution the expected value of $x(k)$. Furthermore, let us subtract (1.180) from (1.177) to obtain $x(k+1) - Ex(k+1)$ and then postmultiply both sides of the resultant equation by $(x'(k+1) - Ex'(k+1))$ and then take the expectations of both sides. It follows that

$$P(k+1) = A(k)P(k)A'(k) + B(k)K(k,k)B'(k)$$
$$\qquad + A(k)\{E[x(k) - Ex(k)][u(k) - Eu(k)]'\}B'(k)$$
$$\qquad + B(k)\{E[u(k) - Eu(k)][x(k) - Ex(k)]'\}A'(k) \tag{1.181}$$

where

$$P(k) \triangleq E[x(k) - Ex(k)][x(k) - Ex(k)]'$$

For the special case where $u(k) - m(k)$ is an uncorrelated sequence and since $x(k) - Ex(k)$ is not a function of $u(k)$ the third and fourth terms in the right hand side of (1.181) will be identically zero and consequently equation (1.181) becomes a deterministic (nonlinear) equation whose solution is the desired covariance matrix

$$P(k + 1) = A(k)P(k)A'(k) + B(k)K(k, k)B'(k) \quad (1.182)$$

▶ **example**
Let $x(k)$ and $u(k)$ be scalar random sequences satisfying the equation

$$x(k + 1) = 0.5x(k) + u(k)$$
$$x(0) = 2$$
$$Eu(k) = 1$$
$$E[u(k_1) - 1][u(k_2) - 1] = \Delta(k_2 - k_1)$$

From (1.180) we have

$$Ex(k + 1) = 0.5Ex(k) + 1$$

and from (1.182) it follows that

$$P(k + 1) = 0.25P(k) + 1 \quad (1.183)$$

where

$$P(k) = E[x(k) - Ex(k)]^2$$

Hence

$$P(0) = 0$$

Consequently, in the case of this simple example, (1.182) assumes the simple form given by (1.183) and hence can be easily solved.

$$P(k) = \sum_{i=0}^{k-1} (0.25)^i; \quad k \geq 1$$

and

▶ $$\lim_{k \to \infty} P(k) = \tfrac{4}{3}$$

1.37 Gaussian (Normal) Processes

The scalar-valued random variable x is called normal (or gaussian) if its probability density function $p(x)$ is given by

$$p(x) = \frac{1}{[2\pi E[x - Ex]^2]^{1/2}} \exp\left\{-\frac{1}{2}\frac{[x - Ex]^2}{E[x - Ex]^2}\right\} \quad (1.184)$$

This usually is seen in the form $p(x) = \dfrac{1}{\sigma\sqrt{2\pi}} \exp\left[-\dfrac{1}{2}\dfrac{(x - m_x)^2}{\sigma^2}\right]$.

An n-dimensional random vector x is called gaussian if its probability density function is given by

$$p(x) = \frac{1}{(2\pi)^{n/2} |P|^{1/2}} \exp\left\{-\tfrac{1}{2}[x - Ex]'P^{-1}[x - Ex]\right\} \qquad (1.185)$$

where

$$P = E[x - Ex][x - Ex]'$$

is the covariance matrix of x and

$$|P| = \text{determinant of } P$$

The components of the vector x are said to be "jointly gaussian" random (scalar) variables.

An n-dimensional process $x(t)$ or $x(k)$ is called a gaussian random process if its joint probability density function $p[x(t_1), x(t_2), \ldots, x(t_i)]$ or $p[x(k_1), \ldots, x(k_i)]$ for all finite values of i can be described by a function of the *form* (1.185).

By carrying out a direct but tedious integration, we can verify that if a random vector x has a density function given by (1.184) its mean and covariance functions are as indicated by that equation: namely, Ex and P.

Gaussian random processes (and variables) have a number of properties of fundamental importance, all stemming from the nature of the function given by (1.184). These are stated in the following.

1. The gaussian density function of a random vector x is completely determined by specifying the mean vector and covariance matrix of x. Consequently, if a gaussian random process $x(t)$ is wide sense stationary (meaning that its mean and covariance matrix are independent of time origin), it will also be stationary in the strict sense.

2. Orthogonal gaussian random variables (or processes) are independent. Proof: When components of a vector-valued random variable are orthogonal, by definition, the vector's covariance matrix P is a diagonal matrix:

$$P = E[x - E(x)][x - E(x)]' = \text{diag }[\lambda_1, \ldots, \lambda_n] \qquad (1.186)$$

Consequently, provided none of the λ_i is zero,

$$P^{-1} = \text{diag }[\lambda_1^{-1}, \ldots, \lambda_n^{-1}] \qquad (1.187)$$

Hence (1.5) can be written as

$$p(x) = \prod_{i=1}^{n} C_i \exp\left\{\frac{-\tfrac{1}{2}[x_i - Ex_i]^2}{E[x_i - Ex_i]^2}\right\} \qquad (1.188)$$

where $C_i = \dfrac{1}{\sqrt{2\pi E[x_i - Ex_i]^2}}$. Since the joint (in this case, gaussian) density function $p(x_1, \ldots, x_n)$ is written in the form of products of functions, each a function of only one of the scalar variables x_i, the assertion is proved by definition of independence.

3. Any linear function of jointly gaussian random variables (or processes) results in another gaussian random variable (or process).*
4. A gaussian process is completely determined (statistically) by specifying its first and second order statistics. This is evident since, for example, the terms needed to construct the mth joint density function of, say, a scalar-valued random process $x(t)$ are only the first and second moments of $x(t)$ which are completely specified by having the first order statistics $p[x(t)]$ and the second order statistics $p[x(t_i), x(t_j)]$.

1.38 Markov Processes

A stochastic sequence $x(k)$ is Markov of first order, or simply Markov, if for every i the following relationship holds

$$p\{x(i) \mid x(i - 1), \ldots, x(1)\} = p\{x(i) \mid x(i - 1)\} \qquad (1.189)$$

In other words, the conditional probability density function of $x(i)$ conditioned on all its past (given) values is the same as using the value in the immediate past.

Equation (1.189) is to be satisfied for all i. Hence, it can equivalently be written as

$$p\{x(i) \mid x(k); k \leq i - 1\} = p\{x(i) \mid x(i - 1)\} \qquad (1.190)$$

Similarly a Markov process $x(t)$ is defined by

$$p\{x(t_i) \mid x(\tau); \tau \leq t_{i-1}\} = p\{x(t_i) \mid x(t_{i-1})\}$$

where

$$t_1 < t_2 < t_3 \cdots < t_i$$

The solution to a general first order differential or difference equation with an independent process (usually, an "uncorrelated" gaussian process) as forcing function is a Markov process. That is, if $x(t)$ and $x(k)$ are n-vectors satisfying

$$\dot{x}(t) = f[x(t), t, u(t)] \qquad (1.191)$$

or

$$x(k + 1) = f[x(k), k, u(k)] \qquad (1.192)$$

where $u(t)$ and $u(k)$ are r-dimensional independent random vectors, the solutions $x(t)$ and $x(k)$ are then (vector) Markov processes. The reason is

---

* For a simple proof, *see* Davenport and Root, *An Introduction to the Theory of Random Signals and Noise*.

intuitively obvious since, for example, in

$$p[x(i) \mid x(i-1), \ldots, x(1)] \qquad (1.193)$$

the term $x(i-1)$ alone used as initial conditions in (1.192) would determine the solution $x(k)$, $k > i - 1$ uniquely if $u(k) = 0$, $k \geq i - 1$. Since $u(k)$ is an independent process and $x(k)$ is only a function of $u(i)$, $i \leq k - 1$, it follows that no information concerning $u(k)$ for $k \geq i - 1$ can be obtained by keeping the terms $x(i-2), \ldots, x(1)$ in (1.193). This is only an intuitive argument, and a rigorous proof involves writing the expression for (1.193) by using (1.192) and showing the desired equality (1.190).

One basic simplification introduced by the Markov property is revealed by considering the joint density function

$$p[x(1), x(2), \ldots, x(m)] \qquad (1.194)$$

Using the Bayes' formula it follows that

$$p[x(1), x(2), \ldots, x(m)] = p[x(m) \mid x(m-1), \ldots, x(1)]$$
$$\times\, p[x(m-1) \mid x(m-2), \ldots, x(1)] \cdots p[x(1)] \qquad (1.195)$$

If we use the Markov property, (1.195) can then be written as

$$p[x(1), \ldots, x(m)] = p[x(1)] \prod_{i=2}^{m} p[x(i) \mid x(i-1)] \qquad (1.196)$$

In other words, the expression of mth joint density function can be written as a product of first order conditional density functions. Although by appearance (1.196) does not look simpler than its equivalent form given by (1.194), it will become evident in later chapters that the expression (1.196) yields considerable simplification, especially in computational developments.

In general, if we have a system governed by an nth order differential or difference equation, which is derived by an independent noise process or sequence, respectively, the solution can be made into a Markov process by representing the nth order equation by an n-dimensional first order vector equation as discussed in Section 1.14.

If $x(k)$ is a Markov process, we can easily establish that its conditional expectation has properties similar to the process itself, i.e.,

$$p[x(i) \mid x(i-1), \ldots, x(1)] = p[x(i) \mid x(i-1)]$$

implies

$$E\{x(i) \mid x(i-1), \ldots, x(1)\} = E\{x(i) \mid x(i-1)\} \qquad (1.197)$$

This follows directly from the definition of conditional expectation

$$E\{x(i) \mid x(i-1), \ldots, x(1)\} = \int_{-\infty}^{+\infty} x(i) p\{x(i) \mid x(i-1), \ldots, x(1)\}\, dx(i)$$
$$(1.198)$$

▶ **example**

Let $x(k)$ be an n-dimensional random sequence satisfying the linear difference equation

$$x(k + 1) = Ax(k) + Bu(k) \qquad (1.199)$$

where $u(k)$ is an r-dimensional random vector such that

$$Eu(k) = 0$$

$$p[u(1), u(2), \ldots] = p[u(1)]p[u(2)] \cdots$$

Here, each $p[u(i)]$ is a probability density on the r-vector $u(i)$. The conditional density function of $x(i)$ given $x(i - 1), \ldots$ can be written as

$$p\{x(i) \mid x(i - 1), x(i - 2), x(i - 3), \ldots\}$$

From (1.199) the random vector $x(i)$ given $x(i - 1)$ is seen to have a mean given by $Ax(i - 1)$ and a density function about this mean identical with that of the random vector $Bu(i - 1)$. Hence

$$p[x(i) - Ax(i - 1)] = p[Bu(i - 1)] \qquad (1.200)$$

Now the terms $x(i - 2), x(i - 3), \ldots$ all can be obtained in terms of $u(i - 3), u(i - 4), \ldots$. Therefore, conditioning the random vector $Bu(i)$ in the right hand side of (1.200) on $x(i - 2), \ldots$ does not alter the form of $p[Bu(i)]$, i.e.,

$$p[Bu(i) \mid x(i - 2), \ldots] = p[Bu(i - 1)]$$

Hence

$$p[x(i) - Ax(i - 1) \mid x(i - 2), \ldots] = p[x(i) - Ax(i - 1)]$$

Since

$$p[x(i) \mid Ax(i - 1)] = p[x(i) - Ax(i - 1)]$$

Therefore

$$p[x(i) \mid x(i - 1), \ldots] = p[x(i) \mid x(i - 1)]$$

and

▶ $$E\{x(i) \mid x(i - 1), \ldots x(1)\} = E\{x(i) \mid x(i - 1)\}.$$

FOURIER TRANSFORM
AND SPECTRAL DENSITY

1.39 Fourier Transform

A useful mathematical tool, when dealing with stationary random processes in connection with linear systems, is the Fourier transformation. The Fourier transform of a scalar function $h(t)$, if it exists, is defined by

$$H(j\omega) = \int_{-\infty}^{+\infty} h(t)e^{-j\omega t}\,dt \qquad (1.201)$$

The inverse Fourier transform, if it exists, is given by

$$h(t) = \frac{1}{2\pi} \int_{-\infty}^{+\infty} H(j\omega)e^{j\omega t} \, d\omega \qquad (1.202)$$

A sufficient condition for the integral in (1.201) and (1.202) to exist is that $h(t)$ be absolutely integrable, i.e.,

$$\int_{-\infty}^{+\infty} |h(t)| \, dt < \infty \qquad (1.203)$$

Let us multiply both sides of (1.201) by $H(-j\omega)$ and integrate over all ω. It follows that

$$\int_{-\infty}^{+\infty} H(j\omega)H(-j\omega) \, d\omega = \int_{-\infty}^{+\infty} \left[\int_{-\infty}^{+\infty} h(t)e^{-j\omega t} \, dt \right] H(-j\omega) \, d\omega \qquad (1.204)$$

Interchanging the order of integration in the right hand side of (1.204) and using (1.202) yields

$$\int_{-\infty}^{+\infty} [h(t)]^2 \, dt = \frac{1}{2\pi} \int_{-\infty}^{+\infty} H(j\omega)H(-j\omega) \, d\omega \qquad (1.205)$$

Equation (1.205) is known as Parseval's theorem. Parseval's theorem enables us to express the energy in the waveform $h(t)$ in terms of the Fourier transform $H(\omega)$.

If $g(t)$ is the impulse response of a linear time-invariant system, the input $u(t)$ and output $y(t)$ (both scalar functions) are related by the convolution integral

$$y(t) = \int_{-\infty}^{\infty} g(t - s)u(s) \, ds \qquad (1.206)$$

Taking the Fourier transform of both sides yields

$$Y(j\omega) = \int_{-\infty}^{+\infty} \left[\int_{-\infty}^{\infty} g(t - s)u(s) \, ds \right] e^{-j\omega t} \, dt \qquad (1.207)$$

If the system is causal, i.e., the output $y(t)$ cannot be related to the values of input $u(s)$ for $s > t$, then in (1.206) we must have

$$g(\tau) = 0 \qquad \tau < 0$$

or $\qquad\qquad\qquad\qquad\qquad\qquad\qquad\qquad\qquad\qquad\qquad\qquad$ (1.208)

$$g(t - s) = 0 \qquad t < s$$

Under this restriction,

$$\int_{-\infty}^{t} g(t - s)u(s) \, ds = \int_{-\infty}^{+\infty} g(t - s)u(s) \, ds \qquad (1.209)$$

If we apply the change of variable $t = s + \tau$, (1.207) can be represented equivalently by

$$Y(j\omega) = \int_{-\infty}^{+\infty} g(\tau)e^{-j\omega\tau}\,d\tau \int_{-\infty}^{+\infty} u(s)e^{-j\omega s}\,ds$$

Finally

$$Y(j\omega) = G(j\omega)U(j\omega) \tag{1.210}$$

where $G(j\omega)$ and $U(j\omega)$ are the Fourier transforms of $g(t)$ and $u(t)$ respectively. Note that (1.210) is valid only if functions $u(t)$ and $g(t)$ are both absolutely integrable (i.e., the transforms $G(j\omega)$ and $U(j\omega)$ should exist). This will imply that the output $y(t)$ is absolutely integrable and that $Y(j\omega)$ exists.

We remark that if $h(t)$ is an even function, i.e.,

$$h(t) = h(-t)$$

then from (1.201) and the identity $e^{-j\omega t} \triangleq \cos \omega t - j \sin \omega t$ it follows that

$$H(j\omega) = \int_{-\infty}^{+\infty} h(t) \cos \omega t\,dt = 2 \int_0^{\infty} h(t) \cos \omega t\,dt$$

Since $\cos \omega t$ is an even function of ω, we can see that $H(j\omega)$ is an even function of ω, i.e.,

$$H(j\omega) = H(-j\omega) \tag{1.211}$$

1.40 Spectral Density

Let $y(t)$ be a zero-mean scalar random process with autocorrelation $\varphi_{yy}(\tau)$

$$E[y(t)y(t + \tau)] = \varphi_{yy}(\tau) \tag{1.212}$$

Here, the fact that $y(t)$ has zero mean implies that the autocorrelation is equal to the covariance.

The spectral density $\Phi_{yy}(j\omega)$ is defined as the Fourier transform of $\varphi_{yy}(\tau)$

$$\Phi_{yy}(j\omega) = \int_{-\infty}^{+\infty} \varphi_{yy}(\tau)e^{-j\omega\tau}\,d\tau \tag{1.213}$$

where it is assumed that $\varphi_{yy}(\tau)$ is absolutely integrable

$$\int_{-\infty}^{+\infty} |\varphi_{yy}(\tau)|\,d\tau < \infty \tag{1.214}$$

The inverse transform relation yields

$$\varphi_{yy}(\tau) = \frac{1}{2\pi} \int_{-\infty}^{+\infty} \Phi_{yy}(j\omega)e^{j\omega\tau}\,d\omega$$

Setting $\tau = 0$, we obtain an interesting relationship, namely an expression for the average energy corresponding to $y(t)$

$$Ey^2(t) = \varphi_{yy}(0) = \frac{1}{2\pi} \int_{-\infty}^{+\infty} \Phi_{yy}(j\omega)\,d\omega$$

Since, as pointed out in Section 1.39, $\varphi(\tau)$ is an even function of τ, then the steps leading to (1.211) insure that the spectral density is always an even function of ω:

$$\Phi_{yy}(j\omega) = \Phi_{yy}(-j\omega)$$

This implies that the spectral density is always a real function. Consequently we do not need to show the explicit dependence on j in its argument, and we may use the notation $\Phi_{yy}(\omega)$.

If $y(t)$ and $x(t)$ are two random processes having stationary cross-correlations $\varphi_{yx}(\tau)$ and $\varphi_{xy}(\tau)$ defined by

$$Ey(t)x(t + \tau) = \varphi_{yx}(\tau)$$
$$Ex(t)y(t + \tau) = \varphi_{xy}(\tau)$$

It is easily seen that

$$\varphi_{xy}(\tau) = \varphi_{yx}(-\tau)$$

The subscripts xy and yx are important here, because in general

$$\varphi_{xy}(\tau) \neq \varphi_{xy}(-\tau)$$

and consequently the cross-spectral density (defined as the Fourier transform of the crosscorrelation)

$$\Phi_{xy}(\omega) = \int_{-\infty}^{+\infty} \varphi_{xy}(\tau)e^{-j\omega\tau}\, d\tau$$

may not, in general, be a real function of ω.

▶ **example**

Let $y(t)$ have the autocorrelation function $\varphi(\tau)$ given by

$$E[y(t)y(t + \tau)] = \varphi(\tau) = e^{-|\tau|}$$

Its spectral density is then

$$\Phi_{yy}(\omega) = \int_{-\infty}^{0} e^{+\tau}e^{-j\omega\tau}\, d\tau + \int_{0}^{\infty} e^{-\tau}e^{-j\omega\tau}\, d\tau = \frac{1}{1 - j\omega} + \frac{1}{1 + j\omega} = \frac{2}{1 + \omega^2}$$

The average energy in $y(t)$ is

$$Ey^2(t) = \varphi(0) = 1$$

where it is seen that

$$\frac{1}{2\pi}\int_{-\infty}^{+\infty} \Phi_{yy}(\omega)\, d\omega = \frac{1}{\pi}\int_{0}^{\infty} \frac{2}{1 + \omega^2}\, d\omega = 1$$

1.41 Linear Systems

Let input $u(t)$ and output $y(t)$ of a system be related by expression (1.206), where $u(t)$ is a random process having autocorrelation function

$$\varphi_{uu}(\tau)$$

It is often of interest to determine the relationship between $\Phi_{yy}(\omega)$ and $\Phi_{uu}(\omega)$. From (1.206), and upon interchanging the order of integration and expectation

$$\varphi_{yy}(\tau) = Ey(t)y(t + \tau) = \int_{-\infty}^{t+\tau} \int_{-\infty}^{t} g(t + \tau - \sigma)g(t - s)Eu(\sigma)u(s)\, d\sigma\, ds$$

If $g(t)$ is the impulse response corresponding to a physically realizable system, then $g(t) = 0$ for $t < 0$ and consequently

$$\varphi_{yy}(\tau) = \int_{-\infty}^{+\infty} \int_{-\infty}^{+\infty} g(t + \tau - \sigma)g(t - s)\varphi_{uu}(\sigma - s)\, d\sigma\, ds \quad (1.215)$$

Taking the Fourier transform of both sides of (1.215) yields

$$\Phi_{yy}(\omega) = \int_{-\infty}^{+\infty} \int_{-\infty}^{+\infty} \int_{-\infty}^{+\infty} g(t + \tau - \sigma)g(t - s)\varphi_{uu}(\sigma - s)e^{-j\omega\tau}\, d\sigma\, ds\, d\tau$$

$$(1.216)$$

which equivalently can be written as

$$\Phi_{yy}(\omega) = \int_{-\infty}^{+\infty} \int_{-\infty}^{+\infty} \int_{-\infty}^{+\infty} g(t + \tau - \sigma)g(t - s)\varphi_{uu}(\sigma - s)$$

$$\times\, e^{-j\omega(t+\tau-\sigma)}e^{+j\omega(t-s)}e^{-j\omega(\sigma-s)}\, d\sigma\, ds\, d\tau \quad (1.217)$$

Using the triple change of variables indicated below,

$$t + \tau - \sigma = \alpha$$
$$t - s = \beta$$
$$\sigma - s = \gamma$$

it follows that

$$\Phi_{yy}(\omega) = \int\!\!\int\!\!\int_{-\infty}^{+\infty} g(\alpha)g(\beta)\varphi_{uu}(\gamma)e^{-j\omega\alpha}e^{+j\omega\beta}e^{-j\omega\gamma}\, d\alpha\, d\beta\, d\gamma$$

Consequently

$$\Phi_{yy}(\omega) = G(j\omega)G(-j\omega)\Phi_{uu}(\omega) \quad (1.218)$$

Similarly, we can find expressions for the input-output spectral density $\Phi_{uy}(\omega)$. We have

$$Eu(t)y(t + \tau) = \varphi_{uy}(\tau)$$

Substituting for $y(t)$ from (1.206) $\big(g(t) = 0, \, t < 0\big)$, it follows that

$$\varphi_{uy}(\tau) = E \int_{-\infty}^{+\infty} g(t + \tau - s)u(s)u(t)\,ds$$

or

$$\varphi_{uy}(\tau) = \int_{-\infty}^{+\infty} g(t + \tau - s)\varphi_{uu}(s - t)\,ds \qquad (1.219)$$

Taking the transform of both sides and using $g(t) = 0, \, t \leq 0$, yields

$$\Phi_{uy}(j\omega) = \int_{-\infty}^{+\infty}\int_{-\infty}^{+\infty} g(t + \tau - s)\varphi_{uu}(s - t)e^{-j\omega\tau}\,ds\,d\tau \qquad (1.220)$$

which can be written equivalently as

$$\Phi_{uy}(j\omega) = \int_{-\infty}^{+\infty}\int_{-\infty}^{+\infty} g(t + \tau - s)\varphi_{uu}(s - t)e^{-j\omega(t+\tau-s)}e^{-j\omega(s-t)}\,ds\,d\tau \qquad (1.221)$$

Using the dual change of variables

$$t + \tau - s = \alpha$$
$$s - t = \beta$$

equation (1.221) becomes

$$\Phi_{uy}(j\omega) = \int_{-\infty}^{+\infty}\int_{-\infty}^{+\infty} g(\alpha)e^{-j\omega\alpha}\varphi_{uu}(\beta)e^{-j\omega\beta}\,d\alpha\,d\beta$$

or finally

$$\Phi_{uy}(j\omega) = G(j\omega)\Phi_{uu}(\omega) \qquad (1.222)$$

Similarly we can show that

$$\Phi_{yu}(-j\omega) = G(j\omega)\Phi_{uu}(-j\omega) \qquad (1.223)$$

as was mentioned in the previous discussion on cross correlations.

▶ **example**

Let the impulse response of a linear, constant-coefficient system be given by

$$g(t) = e^{-t} \quad t \geq 0$$
$$= 0 \quad t < 0$$

Let the input $u(t)$ be a zero-mean random process with autocorrelation function given by

$$\varphi_{uu}(\tau) = e^{-2|\tau|}$$

and let the output $y(t)$ be given by the convolution (1.206). Transforming $g(t)$ and φ_{uu}, we have

$$G(j\omega) = \frac{1}{1 + j\omega}$$

and

$$\Phi_{uu}(j\omega) = \frac{1}{1 - 2j\omega} + \frac{1}{1 + 2j\omega} = \frac{2}{1 + 4\omega^2}$$

Hence, from (1.218)

$$\Phi_{yy}(j\omega) = \frac{1}{1 + \omega^2} \cdot \frac{2}{1 + 4\omega^2}$$

and from (1.222) it follows that

$$\Phi_{uy}(j\omega) = \frac{1}{1 + j\omega} \cdot \frac{2}{1 + 4\omega^2}$$

and using (1.223)

$$\Phi_{yu}(-j\omega) = \Phi_{uy}(j\omega) = \frac{1}{1 + j\omega} \cdot \frac{2}{1 + 4\omega^2}$$

so that, if we wish to eliminate the minus sign from the argument in $\Phi_{yu}(-j\omega)$, we obtain

$$\Phi_{yu}(j\omega) = \frac{1}{1 - j\omega} \cdot \frac{2}{1 + 4\omega^2}$$

Thus, we have derived the Fourier transforms of the system output correlation and the input-output crosscorrelations from knowledge of the ▶ input autocorrelation and the system impulse response.

CALCULUS
OF VARIATIONS

1.42 Derivation of a Necessary Condition

A mathematical tool for determining the location of an extremum (i.e., a point at which a relative maximum or minimum occurs) of a smooth, scalar-valued function of, say, n variables

$$f(x) = f(x_1, \ldots, x_n) \qquad (1.224)$$

is available through an application of the differential calculus. A necessary condition for the function $f(x)$ to have an extremum at the point whose coordinates are $(\hat{x}_1, \hat{x}_2, \ldots, \hat{x}_n)$ is that the equation

$$\operatorname{grad} f(x)\big|_{x=\hat{x}} = 0 \qquad (1.225)$$

be satisfied. The implication here is that if $f(x)$ has an extremum at $(\hat{x}_1, \hat{x}_2, \ldots, \hat{x}_n)$, this implies that (1.225) is satisfied. The gradient of $f(x)$

was defined in Section 1.12 as

$$\operatorname{grad} f(x) \triangleq \frac{\partial f}{\partial x} \triangleq \begin{bmatrix} \dfrac{\partial f(x)}{\partial x_1} \\[2mm] \dfrac{\partial f(x)}{\partial x_2} \\ \cdot \\ \cdot \\ \cdot \\ \dfrac{\partial f(x)}{\partial x_n} \end{bmatrix} \tag{1.226}$$

Let us now consider a "functional," defined as a scalar-valued function of another function. For example, if $x(t)$ is any scalar function defined on the interval $[0, T]$, a functional of the function $x(t)$ might be the area under the graph of $x(t)$:

$$J(x) = \int_0^T x(t)\, dt \tag{1.227}$$

A functional J, such as that in (1.227), is defined for all functions $x(t)$ belonging to some class of functions. (In the present example, the value of $J(x)$ is defined for any function $x(t)$ whose integral exists.) Suppose we are searching for a particular function $x(t)$ denoted by $\hat{x}(t)$ such that

$$J(\hat{x}) \leq J(x) \tag{1.228}$$

or

$$J(\hat{x}) \geq J(x) \tag{1.229}$$

for all functions $x(t)$ belonging to some class of functions. Even though such an extremization problem is analogous to the n-variable problem, we cannot use differential calculus to find which function is the extremal function $\hat{x}(t)$, because the number of points needed to specify the function $\hat{x}(t)$ on the interval $[0, T]$ is infinite. Let us discuss (1.228), that is, the minimization problem. The maximization problem is treated identically.

Let us assume that the extremal function $\hat{x}(t)$ exists satisfying (1.228). The right hand side of (1.228) can be written equivalently as

$$J[\hat{x}(t) + y(t)]; \quad \text{all} \quad y(t) \tag{1.230}$$

Hence, the inequality

$$J[\hat{x}] \leq J[\hat{x}(t) + y(t)]; \quad \text{all} \quad y(t) \tag{1.231}$$

is a necessary and sufficient condition that $\hat{x}(t)$ yield the minimum value of J for all possible functions $x(t)$. Clearly if (1.231) is satisfied, then it is "necessary" that

$$J[\hat{x}(t)] \leq J[\hat{x}(t) + \alpha y(t)]; \quad \text{all} \quad y(t) \tag{1.232}$$

where α is an arbitrary scalar constant since the statement of "all $y(t)$" necessarily includes the function $\alpha y(t)$.

The classical calculus of variations concerns itself with finding functions $x(t)$ which extremize a very simple type of functional, namely a definite integral of the form

$$J[x(t)] = \int_{t_1}^{t_2} f[x(t)]\, dt \qquad (1.233)$$

where $f[x]$ is a scalar-valued function of x. Equation (1.232) then becomes

$$\int_{t_1}^{t_2} f[\hat{x}(t)]\, dt \le \int_{t_1}^{t_2} f[\hat{x}(t) + \alpha y(t)]\, dt; \qquad \text{all} \qquad y(t) \qquad (1.234)$$

If $f[x]$ is twice continuously differentiable in x then the integrand in the right hand side of (1.234) can be expanded into a Taylor series in the vicinity of $\alpha = 0$ yielding

$$f[\hat{x}(t) + \alpha y(t)] = f[\hat{x}(t)] + \alpha y'(t)\, \text{grad}\, f[\hat{x}(t)]$$
$$+ \text{ terms in } \alpha \text{ of second order and higher} \qquad (1.235)$$

Here the $\text{grad}\, f[\hat{x}(t)]$ implies the gradient of the scalar-valued function $f[x]$ evaluated at $x = \hat{x}(t)$, and $y'(t)$ is the vector transpose of $y(t)$. Because $f(x)$ was assumed to be twice differentiable, the higher order terms can be ignored if α is chosen small enough. Hence (1.234) becomes

or

$$\int_{t_1}^{t_2} f[\hat{x}(t)]\, dt \le \int_{t_1}^{t_2} f[\hat{x}(t)]\, dt + \alpha \int_{t_1}^{t_2} y'(t)\, \text{grad}\, f[\hat{x}(t)]\, dt \qquad (1.236)$$

$$\alpha \int_{t_1}^{t_2} y'(t)\, \text{grad}\, f[\hat{x}(t)]\, dt \ge 0 \qquad (1.237)$$

Since α has an arbitrary sign, inequality (1.237) can only be satisfied as an equality if

$$\int_{t_1}^{t_2} y'(t)\, \text{grad}\, f[\hat{x}(t)]\, dt = 0 \qquad (1.238)$$

Since elements of $x(t)$ and hence elements of $y(t)$ are arbitrary functions of time, (1.238) can only be satisfied if

$$\text{grad}\, f[\hat{x}(t)] = 0; \qquad t_1 \le t \le t_2 \qquad (1.239)$$

▶ **example**
Let

$$f[x(t)] = [t - x_1(t)]^2 + x_1(t)x_2(t) + [2 - x_2(t)]^2$$

We would like to determine $x_1(t)$ and $x_2(t)$ so that

$$\int_0^1 f[x(t)]\, dt$$

is minimum.

The gradient of $f[x(t)]$, evaluated at $x = x(t)$, is given by

$$\operatorname{grad} f[x(t)] = \begin{bmatrix} -2[t - x_1(t)] + x_2(t) \\ -2[2 - x_2(t)] + x_1(t) \end{bmatrix}$$

Therefore, by setting the gradient equal to zero it follows that

$$x_1(t) = \frac{-4 + 4t}{3}$$

▶
$$x_2(t) = \frac{-2t + 8}{3}$$

1.43   A Special Case

In the above analysis the components of $x(t)$ and hence those of $y(t)$ were assumed to be independent of each other. This condition was utilized to derive expression (1.239) from (1.238).

Let us now consider a special case where $x(t)$ is a two-dimensional vector given by

$$x(t) = \begin{bmatrix} x_1(t) \\ \dot{x}_1(t) \end{bmatrix} \tag{1.240}$$

Furthermore, the initial and terminal values of $x_1(t)$ are assumed known, say zero. Consequently if

$$y(t) = \begin{bmatrix} y_1(t) \\ \dot{y}_1(t) \end{bmatrix} \tag{1.241}$$

then we must set

$$y_1(t_1) = y_1(t_2) = 0 \tag{1.242}$$

From (1.238) it then follows that

$$\int_{t_1}^{t_2} \left\{ y_1(t) \frac{\partial f[\hat{x}_1(t),\, \dot{\hat{x}}_1(t)]}{\partial x_1} + \dot{y}_1(t) \frac{\partial f[\hat{x}_1(t),\, \dot{\hat{x}}_1(t)]}{\partial \dot{x}_1} \right\} dt = 0 \tag{1.243}$$

Using differentiation by parts we have

$$\int_{t_1}^{t_2} \dot{y}_1(t) \frac{\partial f[\hat{x}_1(t),\, \dot{\hat{x}}_1(t)]}{\partial \dot{x}_1}\, dt$$

$$= y_1(t) \frac{\partial f[\hat{x}_1(t),\, \dot{\hat{x}}_1(t)]}{\partial \dot{x}_1} \Bigg|_{t_1}^{t_2} - \int_{t_1}^{t_2} y_1(t) \frac{d}{dt} \frac{\partial f[\hat{x}_1(t),\, \dot{\hat{x}}_1(t)]}{\partial \dot{x}_1}\, dt \tag{1.244}$$

From (1.242) the first term in the right hand side of (1.244) is zero. Substituting (1.244) into (1.243) yields

$$\int_{t_1}^{t_2} y_1(t) \left\{ \frac{\partial f[\hat{x}_1(t), \dot{\hat{x}}_1(t)]}{\partial x_1} - \frac{d}{dt} \frac{\partial f[\hat{x}_1(t), \dot{\hat{x}}_1(t)]}{\partial \dot{x}_1} \right\} dt = 0 \qquad (1.245)$$

Now since $y_1(t)$ is an arbitrary function (1.245) can only be satisfied if the integrand is identically zero

$$\frac{\partial f}{\partial x_1} - \frac{d}{dt} \frac{\partial f}{\partial \dot{x}_1} = 0 \qquad (1.246)$$

Equation (1.246) is the well-known Euler equation of the calculus of variations.

▶ **example**

Let $x(t)$ be a scalar differentiable function such that

$$x(0) = 0; \qquad x(1) = 2 \qquad (1.247)$$

Derive $x(t)$ satisfying these boundary conditions and minimizing

$$J[x(t)] = \int_0^1 [x^2(t) + \dot{x}^2(t)] \, dt \qquad (1.248)$$

The Euler equation (1.246) assumes the form

$$\hat{x}(t) - \frac{d}{dt} \dot{\hat{x}}(t) = 0$$

or

$$-\ddot{\hat{x}}(t) + \hat{x}(t) = 0$$

This second order linear equation has the solution

$$x(t) = C_1 e^{-t} + C_2 e^t$$

The conditions (1.247) imply that

$$C_1 + C_2 = 0$$
$$C_1 e^{-1} + C_2 e = 2$$

Hence

$$C_1 = \frac{2}{e^{-1} - e}$$

and

$$C_2 = \frac{-2}{e^{-1} - e}$$

So that the required function $x(t)$ is given by

▶
$$x(t) = \frac{2}{e^{-1} - e} e^{-t} - \frac{2}{e^{-1} - e} e^{t}$$

1.44 Quadratic Functionals

The condition (1.239) or the Euler equation (1.246) provides merely the necessary conditions* that the value of $x(t)$ minimizing $J[x(t)]$ should satisfy. This is because of the fact that we do not necessarily satisfy (1.228) for all possible functions $x(t)$ but for a much more limited class denoted by $\hat{x}(t) +$ $\alpha y(t)$ with α sufficiently small. However, if there exists only one function $\hat{x}(t)$ which satisfies the conditions (1.239) or (1.246), that function must necessarily be the desired function. In other words, the conditions are then necessary and sufficient. On the other hand, (1.239) and (1.246) are differential equations in terms of $\hat{x}(t)$. If these differential equations become linear, they will, in general, yield unique solutions. Furthermore, it is clear that if $[x(t)]$ is a quadratic function of $x(t)$ (i.e., of the components $x_1(t)$, $x_2(t)$, ...) the gradient in (1.239) and the partial derivatives in (1.246) will be linear in $x(t)$. Consequently, the necessary conditions derived are also sufficient if the function $f[x(t)]$ is a quadratic function.

▶ example
The examples of the preceding two sections both involved quadratic-type functions $f[x(t)]$. It is observed that in both cases unique solutions were obtained. Let us now consider a very simple nonquadratic function f such as

$$f[x(t)] = 1 + \cos [x(t)]$$

and a corresponding functional

$$J(x) = \int_{t_1}^{t_2} f[x(t)]\, dt.$$

It is clear in this case that the minimum of this functional is obtained by choosing $x(t)$ to be a *constant* over the interval $[t_1, t_2]$ equal to

$$x(t) = (2i + 1)\pi, \qquad i \text{ any integer}$$

Let us try to arrive at the same result by using (1.239). We get

$$\sin [x(t)] = 0$$

* Mathematically, a "necessary condition" is understood as follows: If an extremum exists, it will satisfy the necessary condition.

which has the solution

$$x(t) = i\pi, \qquad i \text{ any integer}$$

It is seen that not only we did not arrive at a unique solution, but in fact half of the solutions obtained will minimize and the other half will maximize the functional given by

$$\int_{t_1}^{t_2} f[x(t)]\, dt$$

Thus, application of (1.239) or (1.246) will not of itself guarantee that we ▶ have found the answer to our minimization problem.

PROBLEMS

1. If A and B are $n \times n$ diagonal matrices, show that

$$AB = BA$$

2. Show that the symmetric matrix

$$A = \begin{bmatrix} \alpha & \gamma \\ \gamma & \beta \end{bmatrix}$$

can be transformed to a diagonal matrix B

$$B = TAT'$$

where

$$T = \begin{bmatrix} \cos\theta & \sin\theta \\ -\sin\theta & \cos\theta \end{bmatrix}; \qquad \tan(2\theta) = \frac{2\gamma}{\alpha - \beta}$$

The matrix T defines "Jacobi's transformation."

3. Show that if $x \neq 0$ is an n-dimensional vector, then the $n \times n$ matrix

$$A = xx'$$

has rank 1.

4. Given the quadratic form

$$Q(x) = x_1^2 + x_1 x_2 + x_2^2$$

determine the matrices M and Λ and vector y where

$$y = Mx$$
$$Q(x) = y'\Lambda y$$
$$\Lambda = \text{a diagonal matrix}$$

Comment on whether M is singular or not.

5. Show that products AB and BA of a matrix A of rank m with a nonsingular matrix B have rank m.

6. If A and B are $n \times n$ matrices, show that

$$\operatorname{tr} AB = \operatorname{tr} BA$$

7. Show that interchanging any two rows (or columns) of a matrix or multiplication of the elements of any row (or column) of the matrix by the same nonzero number does not change the rank of the matrix.

8. Prove that the product of two orthogonal matrices is an orthogonal matrix. Prove that the inverse of an orthogonal matrix is an orthogonal matrix.

9. Let A be an $n \times n$ positive definite matrix and B an $n \times n$ singular matrix (i.e., the rank of B is less than n). Show that the matrix BAB' is positive semidefinite.

10. Let C be a $1 \times n$ row vector and A an $n \times n$ constant matrix. The square matrix B is defined by

$$B = \begin{bmatrix} C \\ CA \\ CA^2 \\ . \\ . \\ . \\ CA^{n-1} \end{bmatrix}$$

Show that if $B'B$ is positive definite, it implies that the matrix

$$\sum_{i=0}^{n-1} A^{i'} C' C A^i$$

is positive definite ($A^0 = I$, $A^1 = A$, $A^2 = AA$, etc.).

11. Derive the solution of the vector differential equation

$$\dot{x}(t) = Ax(t) + Bu(t), \qquad x(0) = \begin{bmatrix} 0 \\ 1 \end{bmatrix}$$

$$A = \begin{bmatrix} -1 & 1 \\ 0 & -2 \end{bmatrix}$$

$$B = \begin{bmatrix} 0 \\ 1 \end{bmatrix}$$

$$u(t) = 1 \qquad t \geq 0$$

12. Derive the values of the solution of the vector differential equation

$$\dot{x}(t) = Ax(t) + Bu(t)$$

at $t = iT$, $i = 1, 2, \ldots$ sec.

$$A = \begin{bmatrix} 0 & 5 \\ 0 & -1 \end{bmatrix}, \quad B = \begin{bmatrix} 0 \\ 1 \end{bmatrix}, \quad x(0) = \begin{bmatrix} 0 \\ 1 \end{bmatrix}$$

$u(t) = i \quad iT \leq t < (i+1)T, \quad T = 1 \text{ sec.}, \quad i = 0, 1, 2, \ldots$

13. Find the impulse response $g(t)$ of a system with input $u(t)$ and output $y(t)$ where

$$\dot{x}(t) = Ax(t) + Bu(t)$$
$$y(t) = Cx(t)$$

$$A = \begin{bmatrix} 0 & 1 \\ -3 & -2 \end{bmatrix}, \quad B = \begin{bmatrix} 0 \\ 1 \end{bmatrix}$$

$$C = [1 \quad 1]$$

14. Given the vector difference equation

$$x(k+1) = A(k)x(k)$$

Show that if A is nonsingular, the matrix

$$\prod_{j=i}^{0} A(j) = A(i)A(i-1) \ldots A(0)$$

has properties similar to those of the transition matrix of a differential equation described in Section 1.17.

15. Let x be a scalar random variable with finite mean Ex and finite second moment Ex^2, and let α be a constant. Determine the value of α which minimizes $E[(x - \alpha)^2]$.

16. Let x be a scalar random variable with values limited to the interval (α, β), i.e., $p(x) = 0$ for $x > \beta$ and $x < \alpha$ where $p(x)$ is the probability density of x. Show that

$$\alpha \leq Ex \leq \beta$$

$$E[x - Ex]^2 \leq \frac{(\beta - \alpha)^2}{4}$$

Hint: Utilize the fact that the probability density $p(x)$ must satisfy

$$\int_{\alpha}^{\beta} p(x)\, dx = 1.$$

17. Let a random variable y be given by

$$y = \sum_{k=1}^{k} x_k$$

where x_k, $k = 1, 2, \ldots$ is a random sequence. Obtain the mean and variance of y: (a) if the x_k are uncorrelated and (b) if the x_k are correlated, e.g., $Ex_i x_j = k(i, j)$.

18. Let a and θ be statistically independent random variables where $p(\theta)$ is uniform over the interval $[0, 2\pi]$. The random process $y(t)$ is defined by

$$y(t) = a \cos (t + \theta)$$

Derive an expression for the autocorrelation of this process. Show that

$$Ey(t) = \lim_{T \to \infty} E_t y(t)$$

19. A random process $x(t)$ is defined by the matrix differential equation

$$\dot{x}(t) = Ax(t) + Bu(t); \qquad x(0) = 0$$
$$Eu(t) = 1$$
$$Eu(t_1)u(t_2) = \delta(t_2 - t_1)$$

$$A = \begin{bmatrix} 3 & 2 \\ 2 & 3 \end{bmatrix}; \qquad \begin{bmatrix} 0 \\ 1 \end{bmatrix}$$

Derive expressions for the mean and covariance matrix of $x(t)$, $t \geq 0$.

20. A random sequence $x(k)$ is defined by

$$x(k + 1) = Ax(k) + Bu(k), \qquad x(0) = 0$$
$$Eu(k) = 0$$
$$Eu(k_1)u(k_2) = \Delta(k_2 - k_1)$$

$$A = \begin{bmatrix} 1 & 1 \\ 0 & 1 \end{bmatrix}, \qquad B = \begin{bmatrix} 1 \\ 1 \end{bmatrix}$$

Derive an expression for the mean and covariance matrix $Ex(k)x'(k)$.

21. Let $x(t) = \alpha + n(t)$ be a random process where α is an unknown constant and

$$En(t) = 0, \qquad En(t)n(t + \tau) = \varphi_n(\tau).$$

Introducing the term $\hat{\alpha}$ by

$$\hat{\alpha} = \frac{1}{T} \int_0^T x(t) \, dt$$

show that

$$E\hat{\alpha} = \alpha$$

$$E(\alpha - \hat{\alpha})^2 = \frac{1}{T} \int_{-T}^{+T} \left(1 - \frac{|\tau|}{T} \right) \varphi(\tau)\, d\tau$$

22. Let $x(t)$ have zero-mean and autocorrelation function given by $\varphi_{xx}(\tau)$. Show that the autocorrelation of $y(t) = \dot{x}(t)$ is given by

$$\varphi_{yy}(\tau) = -\frac{d^2 \varphi_{xx}(\tau)}{d\tau^2}$$

23. The autocorrelation of a random process $x(t)$ is given by $\varphi_{yy}(\tau) = e^{-\alpha\tau^2}$. Derive an expression for the spectral density $\phi_{yy}(\omega)$.

24. Let $y(t) = x(t + \alpha) - x(t - \alpha)$ where $x(t)$ is a zero-mean random process with spectral density $\phi_{xx}(\omega)$. Show that

$$\phi_{yy}(\omega) = 4\phi_{xx}(\omega) \sin^2 \alpha\omega$$

Hint: Derive an expression for $\varphi_{yy}(\tau)$ first.

25. It is desired to determine function $y(t)$ which minimizes the integral

$$I = \int_0^1 (1 + \dot{y}^2(t))\, dt$$

and satisfies the end conditions

$$y(0) = 0, \qquad y(1) = 1$$

(a) Show that the desired solution is $y(t) = t$.
(b) Let us set $y(t) = t + z(t)$. Show that in terms of $z(t)$ the integral takes the form

$$I = 2 + \int_0^1 \dot{z}(t)\, dt, \qquad z(0) = z(1) = 0$$

and that the function $z(t)$ which minimizes this integral is $z(t) = 0$.

SELECTED READINGS

Matrix Algebra

Bellman, R., *Introduction to Matrix Analysis*, McGraw-Hill, New York, 1960.
Indritz, J., *Methods in Analysis*, Macmillan, New York, 1963.
Shilov, G. E., *An Introduction to the Theory of Linear Spaces*, Prentice-Hall, Englewood Cliffs, N.J., 1961.

Differential and Difference Equations

Bellman, R., *Stability Theory of Differential Equations*, McGraw-Hill, New York, 1954.

Coddington, E. A. and N. Levinson, *Theory of Ordinary Differential Equations*, McGraw-Hill, New York, 1955.

Pontryagin, L. S., *Ordinary Differential Equations*, Addison-Wesley, Reading, Mass., 1962.

Probability Theory

Battin, R. H. and J. L. Laning, *Random Processes in Automatic Control*, McGraw-Hill, New York, 1956.

Cramér, H., *Mathematical Methods of Statistics*, Princeton University Press, Princeton, N.J., 1946.

Davenport, W. B. and W. L. Root, *An Introduction to the Theory of Random Signals and Noise*, McGraw-Hill, New York, 1958.

Feller, W., *Introduction to Probability Theory*, Wiley, New York, 1953.

Middleton, D., *An Introduction to Statistical Communication Theory*, McGraw-Hill, New York, 1960.

Papoulis, A., *Probability, Random Variables, and Stochastic Processes*, McGraw-Hill, New York, 1965.

Transform Theory

Churchill, R. V., *Fourier Series and Boundary Value Problems*, McGraw-Hill, New York, 1941.

Truxal, John G., *Automatic Feedback Control System Synthesis*, McGraw-Hill, New York, 1955.

Wiener, N., *The Fourier Integral and Certain of Its Applications*, Cambridge University Press, New York, 1933.

Calculus of Variations

Akhiezer, N. I., *The Calculus of Variations*, Blaisdell Publishing Co., New York, 1962.

Gelfand, I. M. and S. V. Fomin, *Calculus of Variations*, Prentice-Hall, Englewood Cliffs, N.J., 1963.

Hildebrand, F. B., *Methods of Applied Mathematics*, Prentice-Hall, Englewood Cliffs, N.J., 1965.

chapter 2
ESTIMATION THEORY CONCEPTS AND CRITERIA

2.1 Introduction

Stochastic estimation is the operation of assigning a value to an unknown system state* or parameter based on noise-corrupted observations involving some function of the state or the parameter. The noise is assumed to have known statistical properties. The value assigned is called the estimate; the system or function of the observations yielding the estimate is called the estimator. We remark that *any* function which assigns an estimate to each observation is an estimator regardless of whether the resulting estimate is close to or far from the "correct" value. The estimation operation is called optimal if the assignment of an estimate is in accordance with minimization of some estimation criterion, or "cost function." In many applications it is meaningful to assign a cost to an estimate representing a quantitative measure of how "good" an estimate is. This cost should then be a function of the estimation error, i.e., the difference between the true value and the estimated value. An optimal estimate is a function of the received observations which is chosen so as to minimize the expected value of the cost function. An estimator which gives such an optimal estimate is

---

* By system state is meant the solution of a dynamic system (generally stochastic) at a given time. The usefulness of dynamic systems as mathematical models for random processes will be discussed in detail in the next chapter.

called a Bayes estimator. A basic feature of a Bayes estimator is that it requires knowledge of an *a priori* probability density function (sometimes referred to as a "prior probability density" to be more brief) which gives the statistical properties of the parameters or states to be estimated. A model of an estimation process involving the states of the dynamic system presented in equation (1.177) in Chapter 1 with the observation as given by $y(t) = Cx(t) + Dv(t)$ (C and D are known matrices), falls within this category when prior density functions for the input $u(t)$ and the observation noise $v(t)$ are assumed known and it is desired to estimate the state $x(t)$ at some time $t = t_1$. There are also many useful estimators which are not necessarily Bayesian. Such estimators can usually be identified by the fact that they do not require the knowledge of the *a priori* density functions for the parameters in question.

This chapter is devoted to an introduction of various types of estimators, Bayesian and others, and their associated estimation criteria. The purpose of this chapter is to give a general study of estimation criteria and some basic properties of various estimates. The derivation of specialized types of estimators (some of which have traditionally been called "filters") will be treated in following chapters.

2.2 Criteria of Estimation

Let us denote the system state or parameter to be estimated by x (scalar- or vector-valued), the observed signal over some interval by Y, and the *a priori* density of x (if it is available) by $p(x)$. For example, x may be the unknown mean of a random process $y(t)$ observed over the interval $0 \leq t \leq t_1$. As a further example, in terms of a differential equation model given by $\dot{x}(t) = x(t) + u(t)$ and the observation $y(t) = x(t) + v(t)$, x may be the value of the solution at some time $t = t_2$ $(x = x(t_2))$; Y may be the observed signal $y(t)$; $0 \leq t \leq t_1$, and $p(x)$ is specified by prior probability densities on random process $u(t)$ and random variable $x(0)$.*

After the observation is made over an interval, for example $y(t)$ over the interval $[0, t_1]$, the Bayes estimation philosophy assumes that the entire information available to the estimator is contained in the *a posteriori* probability density of x which is defined by the conditional probability density function $p(x \mid Y)$ since only the knowledge of this density function is required to estimate a value for the cost "function." Various criterion ("cost") functions, in conjunction with posterior density functions, lead to various types of Bayes estimates of x. In general, one can state that a large number of optimal estimation criteria depend on various significant features of this and other relevant probability density functions.

* The derivation of $p(x)$ from the given information generally is not an easy task. It is merely stated here that such a function exists and is defined based on the given information. The initial condition $x(0)$ either is a given constant or has a known prior density function.

A list of commonly used criteria, Bayesian and non-Bayesian, is given below.

I. *Minimization of Average Cost.* Consider an estimator $\hat{x}(Y)$ which assigns to every value of the observation Y an estimate $\hat{x}$ of the parameter x. The error in estimation is by definition the quantity $x - \hat{x}$. It is reasonable to associate with each estimate a cost denoted by $\varphi(x - \hat{x})$ as a non-negative scalar function of the estimation error. The average cost incurred by using the estimator x_{est} is then

$$\text{ave. cost} = E\varphi(x - x_{est}) \mid Y = \int_x \varphi(x - x_{est}) p(x \mid Y)\, dx$$

where $\int_x$ denotes integration over the possible values of x. An optimal estimate $\hat{x}$ is defined as the value for x_{est} for which the average cost is minimum. A special case of the cost function, namely, a quadratic function of $(x - x_{est})$ is often used and it leads to a Bayes "min mean-square estimate." If the cost function $\varphi(x - \hat{x})$ is the absolute error $|x - \hat{x}|$, we could ask for the Bayes estimator which minimizes the expected absolute error.

II. *Minimization of Maximum Absolute Error (Min-Max Error).* In this case the only information extracted from $p(x \mid Y)$ is the range of values of x where $p(x \mid Y) \neq 0$. If this range is finite, then the optimum estimate is defined by

$$\min_{x_{est}} \max_x |(x - x_{est})| = \max_x |(x - \hat{x})| \qquad (2.1)$$

Consequently, the optimal estimate $\hat{x}$ assumes a value midway between the largest and smallest possible value of x as determined by $p(x \mid Y)$. Clearly this estimate suffers from the fact that events with small and large posterior probabilities of occurrence are treated alike. This estimate can be interpreted from a Bayesian point of view by recognizing that it can equivalently be written as

$$\min_{x_{est}} \int_x \left[\max_x \varphi(x - x_{est}) \right] p(x \mid Y)\, dx \qquad (2.2)$$

for any $\varphi(x - x_{est})$ which is a monotonic function of $|x - x_{est}|$. Clearly the min-max estimate (2.1) is meaningful when x is a scalar quantity. On the other hand (2.2), which is equivalent to (2.1) when x is scalar, has meaning for vector-valued parameter x. Recall that the cost function $\varphi(x - x_{est})$ is assumed scalar by definition. We remark that an estimator which minimizes the maximum error is not to be confused with a Bayes estimator which minimizes the *expected value* of the absolute error.

III. *Selection of Mode of $p(x \mid Y)$.* If $p(x \mid Y)$ is the *a posteriori* probability density of x, given the observation Y, a reasonable estimate of x would be to use the mode of this density as the optimal estimate. If the *a priori* probability of x, namely $p(x)$, is uniform, then from Bayes' formula

$$p(x \mid Y^0) = \frac{p(Y^0 \mid x)p(x)}{p(Y^0)} \qquad (2.3)$$

where Y^0 is a given observation, the mode of $p(x \mid Y^0)$ coincides with the maximum of $p(Y^0 \mid x)$.* The mode of $p(Y^0 \mid x)$ is the classical conditional maximum likelihood estimate. The conditional probability density function $p(Y \mid x)$ is quite often referred to as the likelihood function. The *a posteriori* mode estimate can be thought of as having a nature similar to a Bayesian estimate since it is a function defined based on the *a posteriori* density.

IV. *Selection of Posterior Median.* We may choose as an estimate of x its median obtained from the density function, $p(x \mid Y^0)$. However, by definition of median this gives the minimum of expected absolute error, and is a Bayesian estimate which was considered in I above.

The above estimates are shown in Figure 2.1 for a hypothetical probability density $p(x \mid Y)$

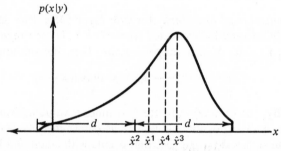

figure 2.1

where $\hat{x}^1$, $\hat{x}^2$, $\hat{x}^3$, and $\hat{x}^4$ are estimates of x under the criteria (1) minimum mean quadratic error, (2) minimax of error, (3) *a posteriori* mode (equivalent to the maximum likelihood estimate if $p(x)$ is uniform), and (4) *a posteriori* median.

V. *Conditional Maximum Likelihood.* After the observation $Y = Y^0$ has been received, a reasonable estimate of the parameter x is to choose that value of the x which gives this particular observation Y^0 a high probability of occurrence. To do this, we take the conditional probability density (likelihood function) $p(Y \mid x)$ and substitute the particular observation Y^0 into it, to obtain $p(Y^0 \mid x)$, which is now a function of x alone. We find the value of

* For any observation $Y = Y^0$, $p(Y^0)$ is a number. Clearly $p(x \mid Y^0)$ and $p[Y^0 \mid x]$ are functions of x.

x which gives the likelihood function its maximum value. This value of x is called the conditional maximum likelihood estimate of x. Thus, the conditional maximum likelihood estimate of x is that value of x which maximizes $p(Y \mid x)$ for the particular observation Y^0 which has been received:

$$\max_x p(Y^0 \mid x) = p(Y^0 \mid \hat{x}) \qquad (2.4)$$

There is a connection between the conditional maximum likelihood and the *a posteriori* mode estimate of x when x happens to have a uniformly distributed prior probability density $p(x)$. To see this, we use the identity

$$p(Y^0 \mid x)p(x) \triangleq p(Y^0, x) \triangleq p(x \mid Y^0)p(Y^0) \qquad (2.5)$$

We recognize that $p(Y^0)$ is just a number (independent of any choice for the estimate of x after the observation is made). Now if $p(x)$ is uniform, the maximum (in x) of $p(Y^0 \mid x)$ must necessarily occur at the same value of x which maximizes $p(x \mid Y^0)$, i.e., the maximum (in x) of $p(Y^0 \mid x)$ and the posterior mode of $p(x \mid Y^0)$ coincide. Therefore, as far as the estimate of x in this case is concerned, the conditional maximum likelihood and the *a posteriori* mode estimates are identical. However, the mechanism of the estimators may be quite different depending on whether $p(Y \mid x)$ or $p(x \mid Y)$ is the available information.

VI. *Unconditional Maximum Likelihood.* In this case, the joint probability density $p(Y, x)$ must be known. Equivalently, the *a priori* probability density $p(x)$ on x, together with the conditional density $p(Y \mid x)$, is required, since

$$p(Y, x) \triangleq p(Y \mid x)p(x) = p(x \mid Y)p(Y)$$

A reasonable estimate is to choose that value of x which maximizes the joint density function $p(Y, x)$ evaluated at any particular observation $Y = Y^0$

$$p(Y^0, \hat{x}) = \max_x p(Y^0, x) = \max_x p(x \mid Y^0)p(Y^0) \qquad (2.6)$$

Thus the unconditional maximum likelihood estimate is the same as the posterior mode estimate introduced in III (above), even though a different density function is maximized.

It has been traditional, in statistical literature, to reserve the term "maximum likelihood" for the *conditional* maximum likelihood estimator. Therefore, in the sequel, the term maximum likelihood estimate will be used only to imply conditional maximum likelihood. When the unconditional estimator is desired, the designation UCML will be used.

VII. *Simultaneous Solution of Moment Equations.* In this case, an appropriate form for the conditional density function $p(Y \mid x)$ is assumed. This density function will contain parameters which adjust the density function to the problem at hand. For example, it could be assumed that the observations Y are distributed according to the gaussian probability density function

$$p(Y \mid x) = \frac{1}{\sqrt{2\pi}\sigma} \exp\left\{ -\frac{1}{2\sigma^2} [Y - x]^2 \right\}$$

where x, which corresponds to the mean of the gaussian density, is an unknown parameter which is to be estimated from the observations Y. The various moments of the observation Y can be computed (using the observation directly) and equated to corresponding values for the moments of the density $p(Y \mid x)$. Consideration of each moment yields one equation involving the unknown x. If x is n-dimensional, then n moments are considered and the resultant n simultaneous equations are solved for x. In the gaussian density example above, if one scalar observation Y^0 were observed, the mean of the observations would also be Y^0. This mean is, of course, the first moment of the observations. Therefore Y^0 is an estimate of the unknown mean x, by the method of moments.

2.3   An Example

The following simple example illustrates the various types of estimates which could be constructed from the same statistical data or probability density functions. Let the parameter to be estimated be a scalar denoted by x with *a priori* probability density function $p(x) = f_x(x)$ where

$$f_x(x) = 1 \qquad 0 \leq x \leq 1$$
$$= 0 \qquad x < 0, x > 1 \qquad (2.7)$$

Let the scalar observation be given by

$$y = x + v \qquad (2.8)$$

which represents "observation noise" having *a priori* probability $p(v) = f_x(v)$

$$f_v(v) = v \qquad 0 \leq v \leq 2$$
$$= 0 \qquad v < 0, v > 2 \qquad (2.9)$$

Furthermore, let us assume that x and v are independent, i.e., $p(x, v) = p(x)p(v)$. We notice that the random variable v tends to prevent our inferring the true value of x from an observation y.

Suppose we have received a single observation

$$y^0 = 2.5 \qquad (2.10)$$

and would like to estimate x.

In order to evaluate various estimates of x we need to derive the conditional density $p(y \mid x)$, the joint density $p(y, x)$, and the posterior density

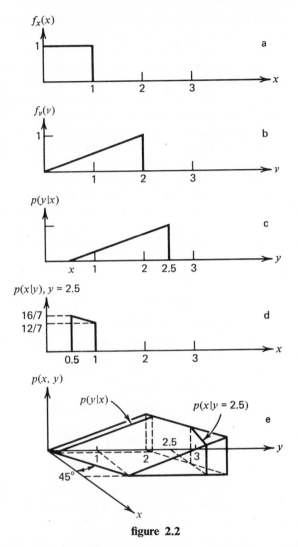

figure 2.2

$p(x \mid y)$. These three densities are derived in the order stated because the statistical data were given as $f_x(x)$ and $f_v(v)$. It will be seen that this order of derivation is a "natural" order in which to proceed. We observe from the relation (2.8) that for a given value of x, the conditional probability density $p(y \mid x)$ is of the same form as $f_v(v)$, except that it is shifted to the left or right by the value of x. For example, $p(y \mid x)$ for $x = 0.5$ is shown in Figure 2.2(c). Now if $p(y \mid x)$ is derived for each x and multiplied by the corresponding value of $f_x(x)$ (in this case $f_x(x) = 1$) the resulting functions may be plotted versus x and y to form the joint probability density function $p(y, x)$, which is shown as the solid figure in Figure 2.2(e). Note that if we were given $p(y, x)$ directly, we could recover $p(y \mid x)$ by taking, for each x, a section of $p(x, y)$ in the y-direction, computing its area (which is the value, for each x, of the marginal density $f_x(x)$) and dividing the height of the section by $f_x(x)$. The resulting section, which is now properly scaled so that its total area (probability) is unity, is $p(y \mid x)$ for the given x. The posterior density $p(x \mid y)$ may be determined from $p(x, y)$ by the same method (as for $p(y \mid x)$) except that the roles of y and x are reversed: First, for the given received value of y, a section of $p(x, y)$ is taken in the x-direction. The area of this section is, by definition, the marginal density $p(y)$ on y. Then, if the height of the section is divided by $p(y)$, the resulting section will have total area equal to unity, and will be the posterior density $p(x \mid y)$ for the given y. For a received value $y = 2.5$, the resulting posterior density function is shown in Figure 2.2(d).

From Figure 2.2(d), which represents the posterior density function $p(x \mid y = 2.5)$, the posterior mode is seen to occur at $x = 0.5$, so that the *a posteriori* mode estimate is $\hat{x} = \frac{1}{2}$,

$$\hat{x}_{\text{posterior mode}} = \tfrac{1}{2} \tag{2.11}$$

Now, for $y = 2.5$, we see that if we were to choose $x = \frac{1}{2}$, the right hand corner of $p(y \mid x)$ (which is the largest value of $p(y \mid x)$) would coincide with the observation $y = 2.5$. Thus,

$$\hat{x}_{\text{maximum-likelihood}} = \tfrac{1}{2} \tag{2.12}$$

This result is the same as for the *a posteriori* mode, as was to be expected from the fact that $f_x(x)$ was uniform.

From Figure 2.2(d) we immediately have that the min-max estimate is

$$\hat{x}_{\text{min-max}} = \tfrac{3}{4} \tag{2.13}$$

From Figure 2.2(d), if we define as our Bayes cost function $\varphi(x - \hat{x}) = (x - \hat{x})^2$, we obtain the corresponding Bayes estimate, called a "min mean-square estimate," given by

$$\hat{x}_{\text{min mean-square error}} = \min_{\hat{x}} \int_{1/2}^{1} (x - \hat{x})^2 [-\tfrac{8}{7}x + \tfrac{20}{7}] \, dx \tag{2.14}$$

yielding

$$\hat{x}_{\text{min mean-square error}} = \tfrac{31}{42} \approx 0.738 \tag{2.15}$$

The Bayes estimate which minimizes the expected absolute error is obtained from

$$\min_{\hat{x}} \int_{1/2}^{1} |x - \hat{x}| \, [-\tfrac{8}{7}x + \tfrac{20}{7}] \, dx \tag{2.16}$$

or equivalently

$$\min_{\hat{x}} \left\{ -\int_{1/2}^{\hat{x}} (x - \hat{x})[-\tfrac{8}{7}x + \tfrac{20}{7}] \, dx + \int_{\hat{x}}^{1} (x - \hat{x})[-\tfrac{8}{7}x + \tfrac{20}{7}] \, dx \right\} \quad (2.17)$$

Differentiating with respect to $\hat{x}$ and equating the result to zero yields

$$\int_{1/2}^{\hat{x}} [-\tfrac{8}{7}x + \tfrac{20}{7}] \, dx - \int_{\hat{x}}^{1} [-\tfrac{8}{7}x + \tfrac{20}{7}] \, dx = 0$$

which may be evaluated and solved for $\hat{x}$ to give

$$\hat{x}_{\text{min expected absolute error}} = \frac{5}{2}\left(1 - \frac{\sqrt{2}}{2}\right) \approx 0.732 \quad (2.18)$$

Finally, from $p(y \mid x)$, the first moment of y is given by

$$E_y = \int_x^{x+2} y\left[\tfrac{1}{2}y - \frac{x}{2}\right] dy = x + \tfrac{4}{3}$$

Note that this first moment E_y is a function of the parameter value x. We now observe that the single observation $y = 2.5$ has the statistical average (first moment) equal to 2.5. Hence, solving the equation

$$x + \tfrac{4}{3} = 2.5$$

yields an unacceptable solution for the estimate namely

$$\hat{x}_{\text{method of moments}} = 1.1 \quad (2.19)$$

As is evident here, the method of moments does not always yield acceptable estimates.

2.4 Bayes Minimum Quadratic Cost Estimate

Let the cost function $\varphi(x - x_{\text{est}})$ be a quadratic function (or "quadratic form") of $x - x_{\text{est}}$. The optimal estimate is then obtained by minimizing the cost $(x - x_{\text{est}})'Q(x - x_{\text{est}})$ conditioned on the observation Y where Q is an $n \times n$ symmetric positive definite matrix.

$$E(x - \hat{x})'Q(x - \hat{x}) \mid Y = \min_{x_{\text{est}}} E(x - x_{\text{est}})'Q(x - x_{\text{est}}) \mid Y \quad (2.20)$$

Expanding (2.20) and noticing that $\hat{x}$ is the estimate and therefore a non-random function of the observations, we can obtain $\hat{x}$ by differentiating the right hand side of (2.20) with respect to x_{est}, setting the derivative equal to zero, and solving for the minimizing x_{est} which we shall call $\hat{x}$. We get

$$\hat{x} = Ex \mid Y \quad (2.21)$$

In other words, the optimal estimate of x minimizing the average of a quadratic cost function is the conditional expectation of x given the observation Y. This is a result obtained under a very general setting since neither knowledge of a prior probability density of x nor the form of dependence of Y on x was necessary for the derivation. Such information, however, is required to actually carry out, for each Y, the expectation operation indicated by equation (2.21) to yield a value for $\hat{x}$ as a function of the observation Y. The "estimator" must be designed to carry out this operation. In terms of the differential equation model of Section 2.2, when $y(t)$, $0 \leq t \leq t_1$ is the observed quantity and it is desired to estimate $x(t)$ at $t = t_2$, (2.21) assumes the form

$$\hat{x}(t_2) = Ex(t_2)\big|_{\substack{y(t) \\ 0 \leq t \leq t_1}} \tag{2.22}$$

and the corresponding equation for a similar discrete model* is

$$\hat{x}(k_2) = Ex(k_2)\big|_{y(1), \, \ldots, \, y(k_1)} \tag{2.23}$$

2.5 Gaussian Processes

When all the terms of a vector sequence $x(k)$ are jointly normal with zero-mean, the conditional probability density $p\big(x(k_2) \mid Y\big)$ will also be gaussian if Y is any linear combination of the terms of the sequence (such as the past samples $x(1)$, $x(2)$, ..., $x(k_1)$). In the following it is shown that under the zero-mean, gaussian hypothesis above, the conditional expectation (i.e., the optimal estimator) indicated in (2.23) can be instrumented as a linear combination of the observations $y(1)$, ..., $y(k_1)$. The proof is given for scalar observation $y(i)$ and can simply be extended to the vector case. There is no restriction as to whether $k_1 < k_2$, $k_1 = k_2$, $k_1 > k_2$.

Let us choose vectors $\alpha(i)$, $i = 1$, ..., k_1 to form linear combinations of the observations $y(1)$, $y(2)$, ..., $y(k_1)$, and let us require that the α_i satisfy

$$E\left[x(k_2) - \sum_{i=1}^{k_1} \alpha(i)y(i) \right]y(j) = 0 \qquad j = 1, \ldots, k_1 \tag{2.24}$$

The existence of vectors α_i satisfying (2.24) is guaranteed, since when (2.24) is multiplied out, we get a system of linear equations to be solved for the vectors α_i involving the covariance matrix

$$\text{Cov } [y(i), y(j)]$$

which is presumed nonsingular.

Since, by construction, the vectors $\left[x(k_2) - \sum_{i=1}^{k_1} \alpha(i)y(i) \right]$ and $y(j)$ are jointly gaussian, (2.24) imposes the requirement that they be independent. This

* For example, $x(k + 1) = Ax(k) + u(k)$, $y(k) = Cx(k) + Dv(k)$, $1 \leq k \leq k_1$.

statement is meaningful, since we have established the existence of vectors $\alpha(i)$ satisfying (2.24). Using this independence, it follows that

$$E\left[x(k_2) - \sum_{i=1}^{k_1}\alpha(i)y(i)\right]\Bigg|_{y(1),\ldots,y(k_1)} = E\left[x(k_2) - \sum_{i=1}^{k_1}\alpha(i)y(i)\right] \quad (2.25)$$

and further, by the zero-mean property of the original sequence $x(k)$, it follows that both sides of (2.25) are equal to zero.

Expanding the left hand side of (2.25), and using the fact that the $y(i)$ are known numbers assumed to have been observed, we get

$$Ex(k_2)\big|_{y(1),\ldots,y(k_1)} = \sum_{i=1}^{k_1}\alpha(i)y(i) \quad (2.26)$$

Equation (2.26) is the desired result since the right hand side is a linear function of the observations $y(1),\ldots,y(k_1)$. Thus, we have instrumented the conditional expectation as a linear combination of the observations, and we have provided in (2.24) a means whereby the $\alpha(i)$ can be computed from the covariance matrix Cov $y(i)$, $[y(j)]$, which is presumed known.

Similar results are derived for the random process in connection with (2.22). Let us choose a vector-valued function

$$\alpha(t, \tau) = \begin{bmatrix} \alpha_i(t, \tau) \\ \cdot \\ \cdot \\ \cdot \\ \alpha_n(t, \tau) \end{bmatrix}$$

such that

$$E\left[x(t_2) - \int_0^{t_1}\alpha(t, \tau)y(\tau)\,d\tau\right]y(t) = 0 \quad (2.27)$$

or

$$\int_0^{t_1}\alpha(t, \tau)Ey(t)y(\tau)\,d\tau = Ex(t_2)y(t) \quad (2.28)$$

Equation (2.28) is the well-known Wiener-Kolmogorov equation, which will be discussed later. Again we have

$$E\left[x(t_2) - \int_0^{t_1}\alpha(t, \tau)y(\tau)\,d\tau\right]\Bigg|_{\substack{y(t) \\ 0 \le t \le t_1}} = E\left[x(t_2) - \int_0^{t_1}\alpha(t, \tau)y(\tau)\,d\tau\right] = 0$$

$$(2.29)$$

Hence,

$$Ex(t_2)\big|_{\substack{y(t) \\ 0 \le t \le t_1}} = \int_0^{t_1}\alpha(t, \tau)y(\tau)\,d\tau \quad (2.30)$$

Equation (2.30) indicates the desired result, i.e., the conditional expectation $Ex(t_2)|_{y(t)}$ in terms of a linear function of the observation $y(t)$.

Derivation of conditional expectation as a linear function of observations of gaussian processes is a very significant result, since it is then sufficient to search only among the class of linear systems (filters) operating on the observed data in order to find the optimum (minimum quadratic cost) estimator. Furthermore, since $p(x(k) \mid Y)$ is gaussian, and hence symmetric and unimodal, its *a posteriori* mode coincides with its conditional expectation and with its posterior median.

2.6   Minimization of Average Nonquadratic Cost

It was shown in Section 2.4 that under very general conditions, if the cost is a quadratic function of error, $e = x - \hat{x}$, then the optimal estimate $\hat{x}$ is given by $Ex \mid Y$ where Y is the observation. It will now be shown that when the processes are such that the posterior probability density function $p(x \mid Y)$ is symmetric and unimodal, then the quantity $Ex \mid Y$ is the optimal estimate with respect to a much broader class of cost functions. A case where this condition on $p(x \mid Y)$ is satisfied is when the processes are gaussian.

Let $\varphi(e)$ satisfy the following conditions*

$$\varphi(0) = 0$$
$$\varphi(e_2) \geq \varphi(e_1) \quad \text{for} \quad e_2 \geq e_1 \geq 0 \qquad (2.31)$$
$$\varphi(e) = \varphi(-e)$$

Typical examples of this cost function are shown in Figure 2.3.

The value of the expected cost is given by

$$E\varphi(x - \hat{x}) \mid Y = \int \varphi(x - \hat{x}) p(x \mid Y) \, dx \qquad (2.32)$$

It is desired to choose $\hat{x}$ so that the integral in (2.32) is minimum. The

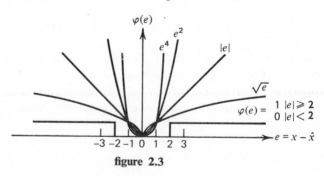

figure 2.3

---

* The term x and consequently $\hat{x}$ and e are assumed scalar, for simplicity in the proofs.

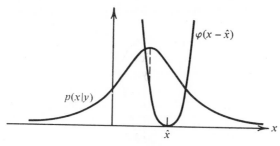

figure 2.4

conditional probability $p(x \mid Y)$ was assumed to be unimodal and symmetric with respect to the mode. The two functions whose product is $\varphi(x - \hat{x})p(x \mid Y)$ are shown in Figure 2.4. The question is: What value of $\hat{x}$ should be selected so that the integral in (2.32) is a minimum.

From this figure it can be deduced that in order to get the minimum value for $E\varphi(x - \hat{x}) \mid Y$ the estimate $\hat{x}$ should coincide with the mode of $p(x \mid Y)$. Since for the assumed posterior probability densities the posterior mode coincides with the conditional expected value, we have

$$\hat{x} = Ex \mid Y$$

The proof of the above statement is simple when $\varphi(e)$ is a differentiable function of e. In order to minimize $E\varphi(x - \hat{x}) \mid Y$ let us differentiate (2.32) with respect to $\hat{x}$ and equate the result to zero

$$\int_{-\infty}^{+\infty} \frac{\partial \varphi(x - \hat{x})}{\partial(x - \hat{x})} p(x \mid Y) \, dx = 0 \tag{2.33}$$

Let $E(x \mid Y) = m$ and using the change of variable $z = x - m$ we have

$$\int_{-\infty}^{+\infty} \frac{\partial \varphi(z + m - \hat{x})}{\partial(z + m - \hat{x})} p[(z + m) \mid Y] \, dz = 0 \tag{2.34}$$

Since $p(x \mid Y)$ is symmetric then $p(x + m) \mid Y$ is an even function of z. Due to the properties of $\varphi(e)$ given by (2.31) the function $\dfrac{\partial \varphi(e)}{\partial(e)}$ is an odd function of e. The integral in (2.34) is now represented by Figure 2.5.

Inasmuch as the product of an odd and an even function is an odd function and the integral of an odd function over symmetric bounds is zero, it is evident that in order to satisfy (2.34) the estimate $\hat{x}$ should be equal to m, which is the desired result. Minor modification of the proof is necessary when $\varphi(e)$ is not differentiable in e.

In fact, for the special case where $p(e \mid Y)$ is gaussian we can generalize the results to consider the case where the cost function is nonsymmetric.

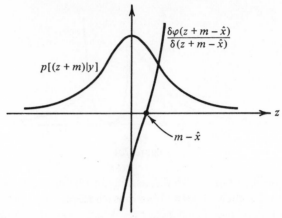

figure 2.5

This is shown here for a scalar-valued x. The generalization to vector-valued x is straightforward. Let us assume that $\varphi(e)$ satisfies the following

$$\varphi(0) = 0$$

$$\varphi(e_2) \geq \varphi(e_1) \quad \text{for} \quad e_2 \geq e_1 \geq 0 \tag{2.35}$$

$$\varphi(e_2) \geq \varphi(e_1) \quad \text{for} \quad e_2 \leq e_1 \leq 0$$

and, furthermore, that the error e by hypothesis, is gaussian with zero-mean. Hence, we have

$$p(e \mid Y) = \frac{1}{\sigma\sqrt{2\pi}} \exp\left[-\frac{1}{2}\frac{e^2}{\sigma^2}\right] \tag{2.36}$$

Therefore

$$E\varphi(e) \mid Y = \frac{1}{\sigma\sqrt{2\pi}} \int_{-\infty}^{+\infty} \varphi(e) \exp\left[-\frac{1}{2}\frac{e^2}{\sigma^2}\right] de \tag{2.37}$$

Let us define two functions $\varphi'(e)$ and $\varphi''(e)$ so that

$$\varphi'(e) = \varphi(e) \quad e > 0$$

$$\varphi'(e) = 0 \quad e \leq 0 \tag{2.38}$$

and

$$\varphi''(e) = \varphi(e) \quad e < 0$$

$$\varphi''(e) = 0 \quad e \geq 0 \tag{2.39}$$

Substituting into (2.37) it follows that

$$E\varphi(e) \mid Y = \frac{1}{\sigma\sqrt{2\pi}} \left\{ \int_0^\infty \varphi'(e) \exp\left[-\frac{1}{2}\frac{e^2}{\sigma^2}\right] de + \int_{-\infty}^0 \varphi''(e) \exp\left[-\frac{1}{2}\frac{e^2}{\sigma^2}\right] de \right\} \tag{2.40}$$

Using the change of variable $e = \sigma\varepsilon$ (2.40) becomes

$$E\varphi(e) \mid Y = \frac{1}{\sqrt{2\pi}} \left\{ \int_0^\infty \varphi'(\sigma\varepsilon) \exp\left[-\tfrac{1}{2}\varepsilon^2\right] d\varepsilon + \int_{-\infty}^0 \varphi''(\sigma\varepsilon) \exp\left[-\tfrac{1}{2}\varepsilon^2\right] d\varepsilon \right\}$$

(2.41)

Consider

$$0 \le \sigma_1 \le \sigma_2$$

From (2.38), (2.39), and (2.35) we conclude that

$$\varphi'(\sigma_2\varepsilon) \ge \varphi'(\sigma_1\varepsilon); \qquad \varepsilon \ge 0$$
$$\varphi''(\sigma_2\varepsilon) \ge \varphi''(\sigma_1\varepsilon); \qquad \varepsilon \le 0$$

(2.42)

From (2.41) and (2.42) it is clear that

$$\frac{1}{\sqrt{2\pi}} \int_{-\infty}^{+\infty} \varphi(\sigma_1\varepsilon) \exp\left[-\tfrac{1}{2}\varepsilon^2\right] d\varepsilon \le \frac{1}{\sqrt{2\pi}} \int_{-\infty}^{+\infty} \varphi(\sigma_2\varepsilon) \exp\left[-\tfrac{1}{2}\varepsilon^2\right] d\varepsilon \quad (2.43)$$

Equation (2.43) implies that $E[\varphi(e) \mid Y]$ when considered as a function of $Ee^2 = \sigma^2$ is a nondecreasing function of σ. Hence minimization of Ee^2 yields the desired minimum for $E\varphi(e) \mid Y$ for all functions $\varphi(e)$ satisfying (2.35).

The results of Sections 2.4, 2.5, and 2.6 can be summarized as follows:

1. If the cost to be minimized is quadratic in e, the optimal estimate is $Ex \mid Y$ regardless of the form of probability density $p(x \mid Y)$, provided the expectation $Ex \mid Y$ exists.
2. If the probability density $p(x \mid Y)$ is gaussian, then the optimal estimator minimizing the expected value of a quadratic cost function is a linear system. Furthermore, this system is optimal with respect to a much larger class of cost functions satisfying (2.35). In other words (when $p(x \mid Y)$ is normal) a linear system minimizing Ee^2 is the best system (linear or nonlinear) for minimizing the average of the general cost function $\varphi(e)$.

2.7 Biased and Unbiased Estimates

A conditional unbiased estimate is one whose expected value is equal to the true value of the quantity being estimated. An unconditional unbiased estimate is one whose expected value is equal to the expected value of the quantity being estimated. Here the estimate is taken to be a random variable being a function of the observations. Therefore, if $\hat{x}$ is a conditional unbiased estimate, then

$$E_Y \hat{x} = \int \hat{x} p(Y \mid x) \, dY = x$$

(2.44)

and if $\hat{x}$ is an unconditional unbiased estimate, we have

$$E_Y\hat{x} = \int \hat{x}p(Y)\,dY = Ex \qquad (2.45)$$

For example, the estimate $Ex \mid Y$ is an unconditional unbiased estimate since

$$E_Y Ex \mid Y = \iint xp(x \mid Y)p(Y)\,dx\,dY = Ex \qquad (2.46)$$

These are clearly desirable features since they indicate that the average of estimates of the same quantity taken over many runs will approach the true or the expected value of the quantity. The biased estimates are those which do not have this property.

PROBLEMS

1. Derive the various estimates obtained in the example of Section 2.3 but for the single observation given by $y = 0.5$, 1, and 2.

2. In the example of Section 2.3, replace the probability density function of the observation noise by

$$\begin{aligned} f_v(v) &= \tfrac{1}{2} \quad & 0 \le v \le 2 \\ &= 0 \quad & v < 0, v > 2 \end{aligned}$$

and derive various estimates of x for $y = 2.5$ and $y = 0.5$.

3. Let the scalar observation y be given by

$$y = x + v$$

where

$$p(v) = \tfrac{1}{2}v, \quad 0 \le v \le 2; \text{ zero otherwise}$$

and

$$p(x) = 2x, \quad 0 \le x \le 1; \text{ zero otherwise}$$

For $y = 2.5$ derive the maximum-likelihood, *a posteriori* mode, and minimum-expected-square estimates of x.

4. Let x and v be independent gaussian random variables with zero-mean and respective variances σ_x^2 and σ_v^2. Assume we have observed

$$y = x + v = 1$$

Derive various estimates of x (maximum likelihood, *a posteriori* mode, etc.).

5. Let x and Y be jointly gaussian random variables and let

$$x = E[x \mid Y]$$

Show that

$$E[\hat{x}^2] \le Ex^2$$

SELECTED READINGS

Brown, A. E., Jr., "Asymmetric Non-Mean-Square Error Criteria," *IRE Transactions on Automatic Control*, Vol. AC-7, 1962.

Deutsch, R., *Estimation Theory*, Prentice-Hall, Englewood Cliffs, N.J., 1965.

Helstrom, C. W., *Statistical Theory of Signal Detection*, Pergamon Press, New York, 1960.

Middleton, D., *An Introduction to Statistical Communication Theory*, McGraw-Hill, New York, 1960.

Raiffa, H. and R. Schlaifer, *Applied Statistical Decision Theory*, Harvard University Press, Cambridge, Mass., 1961.

Sherman, S., "Non-Mean-Square Error Criteria," *IRE Transactions on Information Theory*, Vol. IT-4, 1958.

Sherman, S., "A Theorem on Convex Sets with Applications," *Ann. Math. Statist.*, Vol. 26, 1955.

Wainstein, L. A. and V. D. Zubakov, *Extraction of Signals from Noise*, translation from Russian, Prentice-Hall, Englewood Cliffs, N.J., 1962.

Zadeh, L. A., "What is Optimal," *IRE Transactions on Information Theory*, Vol. 4, 1958.

chapter 3 / MODELING OF STOCHASTIC PROCESSES

3.1 Introduction

A typical estimation process can be described as follows: There is a signal function $s(t)$ and it is desired to estimate the value of this function by processing data $v(t)$ (obtained over an interval $0 \leq t \leq T$) which are related to $s(t)$ through the relation

$$v(t) = f[s(t), n(t), t], \qquad 0 \leq t \leq T$$

where $n(t)$ is a random noise. Before the observation $v(t)$ is made, the only knowledge available concerning $s(t)$ and $n(t)$ are their *a priori* statistical properties. Clearly a "good" estimator should weigh the observations on the basis of this knowledge. Mathematically this *a priori* information is contained in the joint probability density function $p[s(t_1), s(t_2), \ldots, n(t_1), n(t_2), \ldots]$ where $t_1, t_2, \ldots$, are successive time instants. Many of the results obtained in the literature apply when $s(t)$ and $n(t)$ are gaussian processes and hence can be statistically characterized by their first two moments only, namely, the mean values $Es(t)$, $En(t)$, and the correlation functions $Es(t_1)s'(t_2)$, $Es(t_1)n'(t_2)$, and $En(t_1)n'(t_2)$. When the processes are stationary, the spectral densities of $s(t)$, $n(t)$ and the cross-spectral density of $s(t)$ and $n(t)$ contain information equivalent to that contained in the first two moments. Even in the case when the processes are not gaussian, we may only be interested in their first two moments and, therefore, the means and correlation functions could then characterize the process adequately.

Each form of the representation of the *a priori* information discussed above, namely, the joint distribution, the moments, the correlation functions, or the spectral densities, can be used to define a mathematical model of the random processes involved.

A different form of the mathematical model for the random processes is the description by means of differential (or difference in case of random sequences) equations. This representation offers many advantages over the previous ones and will be mainly used in this work. This chapter is devoted to the introduction and critical analysis of this representation.

3.2 Differential Equation Model of Random Processes

The basic justification behind the representation of a random process by a differential equation is that the original source of the process in most practical situations is an independent process (white noise) in actuality or by assumption. For example, when the noise at the output of an electronic amplifier is gaussian with an autocorrelation $\varphi(\tau)$, the original source is most likely a thermal noise which can be assumed to be white gaussian which is then processed through the dynamics of the amplifier to yield the autocorrelation $\varphi(\tau)$ at the output.

The differential equation model of a random process is given by the following equations*

$$\dot{x}(t) = f(x(t), t) + B(t)u(t)$$
$$y(t) = g[x(t), v(t), t]$$
(3.1)

where $x(t)$ is an n-vector, f is an n-vector whose components are functions of $x_1(t), \ldots, x_n(t)$ and t, B is an $n \times n$ matrix, $u(t)$ and $v(t)$ are vector functions of time having dimensions r and q respectively. Each component of $u(t)$ and $v(t)$ is defined to be a zero-mean, uncorrelated process, i.e.,

$$Eu(t) = Ev(t) = 0$$
$$Eu(t_1)u'(t_2) = K(t_2, t_1)\delta(t_2 - t_1)$$
$$Ev(t_1)v'(t_2) = L(t_2, t_1)\delta(t_2 - t_1)$$
$$Eu(t_1)v'(t_2) = M(t_2, t_1)\delta(t_2 - t_1)$$
(3.2)

Quantities K, L, and M are matrices and δ represents the Dirac delta function. The vector function $g[x(t), v(t), t] = g[x_1(t), \ldots, x_n(t), v_1(t), \ldots, v_q(t), t]$ is an s-dimensional vector yielding the s-dimensional process $y(t)$.

* The formal solution to (3.1) when the white noise input $u(t)$ is additive as indicated, can be rigorously justified by Ito's Calculus. The general case when f is a nonlinear function of u is not treated here since it requires rigorous treatment beyond the scope of this book. See Doob, J. L., *Stochastic Processes*, Wiley, 1953.

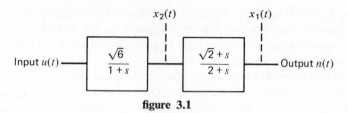

figure 3.1

The advantages of the differential equation model over certain classical models will be discussed in Section 3.4. The relationships between various models are illustrated by means of the following examples.

▶ **example 1**

Let a zero-mean stationary gaussian random process $n(t)$ have the auto-correlation $\varphi(\tau)$ given by

$$\varphi(\tau) = e^{-|\tau|} + e^{-2|\tau|}$$

The spectral density of $n(t)$ is then

$$\phi(\omega) = 6\,\frac{2 + \omega^2}{(1 + \omega^2)(4 + \omega^2)}$$

Using Fourier transform theory, we know that such a random process $n(t)$ can be obtained as the output of a linear system with transfer function $\sqrt{6}\,\dfrac{\sqrt{2} + s}{(1 + s)(2 + s)}$ whose input $u(t)$ is zero-mean white gaussian noise with spectral density unity (Figure 3.1).

Let us introduce the variables $x_1(t)$ and $x_2(t)$ as shown in Figure 3.1 and then notice that the following first order linear differential equations will represent the operations indicated by the transforms in the figure

$$\dot{x}_1(t) + 2x_1(t) = \dot{x}_2(t) + \sqrt{2}x_2(t)$$
$$\dot{x}_2(t) + x_2(t) = \sqrt{6}u(t)$$

After substituting for $\dot{x}_2(t)$ from the second equation and rearranging terms, we can write the differential equations for $x_1(t)$ and $x_2(t)$ in vector form:

$$\dot{x}(t) = \begin{bmatrix} -2 & -1 + \sqrt{2} \\ 0 & -1 \end{bmatrix} x(t) + \begin{bmatrix} \sqrt{6} \\ \sqrt{6} \end{bmatrix} u(t) \tag{3.3}$$

The output $n(t) = x_1(t)$ can be clearly represented by

$$y(t) = [1 \quad 0]x(t) \tag{3.4}$$

We can see that equations (3.3) and (3.4) along with $Eu(t_1)u(t_2) = \delta(t_2 - t_1)$, with $y(t) \triangleq n(t)$ are special cases of (3.1) and (3.2) and represent the process $n(t)$. Notice that no initial condition is given for (3.3). We account for this by regarding (3.3) as giving purely the response to the random input $u(t)$, with the assumption that any response due to initial conditions at $t = -\infty$ has died out. Since $u(t)$ is a stationary process defined over the interval $-\infty \le t \le +\infty$, it is clear that such a steady-state situation is not obtainable in practice if, for example, the aim is to generate $n(t)$ by a computer using (3.3). Two approaches are possible to avoid this difficulty. One is to choose any arbitrary initial condition x_0 at any arbitrary time $t_0(x(t_0) = x_0)$ and recognize that for $t \gg t_0$ (i.e., after running the differential equation for a long time) the output $y(t)$ approaches the stationary process $n(t)$. In other words, we have to wait until the transients in the output have (approximately) died out. The second approach is to choose an initial condition x_0 more carefully to avoid these transients completely, that is, to generate an output $y(t)$, $t \ge t_0$ so that it is (statistically) identical with $n(t)$, $t \ge t_0$. Clearly x_0 as an initial condition would have to be a random vector having the same statistical properties as the solution to (3.3) at $t = t_0$ when $u(t)$, $-\infty \le t \le t_0$ has been applied. Using the derivation in Section 3.6, and evaluating (3.16), we find the initial condition $x(t_0)$ for equation (3.3) will be a zero-mean random vector with covariance matrix given by

$$C[x_1(t_0), x_2(t_0)] = E[x_1(t_0), x_2(t_0)][x_1(t_0), x_2(t_0)]' = \begin{bmatrix} 2 & (1 + \sqrt{2}) \\ (1 + \sqrt{2}) & 3 \end{bmatrix}$$

▶ example 2

Let $n(t)$ be the output of a system with white gaussian input represented by Figure 3.2.

The system includes a nonlinearity described by

$$\beta = \alpha + \alpha^3$$

Although the input $u(t)$ is assumed gaussian the process $n(t)$ will not be gaussian, because of the nonlinearity, and its representation by means of a joint probability density function or various moments assumes a very complicated form. The idea of a differential equation representation is to

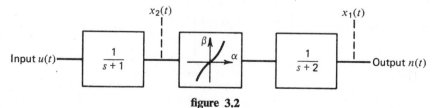

figure 3.2

represent the process as it is actually generated. Again let us introduce the variables $x_1(t)$ and $x_2(t)$ and identify the following two differential equations representing the operations called for by Figure 3.2

$$\begin{aligned} \dot{x}_1(t) &= -2x_1(t) + x_2(t) + x_2{}^3(t) \\ \dot{x}_2(t) &= -x_2(t) + u(t) \end{aligned} \tag{3.5}$$

Equations (3.5) along with

$$y(t) = [1 \ \ 0]x(t)$$

and $Eu(t_1)u(t_2) = \delta(t_2 - t_1)$ represent the process $n(t) = y(t)$. As in Example 1, we assume all transients due to initial conditions have died out, or that random initial conditions having the proper statistics have been ▶ used.

▶ **example 3**
An observed signal $v(t)$ is given by the scalar equation

$$v(t) = s(t) + n(t)$$

where $s(t)$ denotes the signal and is a zero-mean process having spectral density $\phi_{ss}(\omega) = \dfrac{1}{1 + \omega^2}$ and $n(t)$ is additive noise which also has a zero-mean and a spectral density given by $\phi_{nn}(\omega) = \dfrac{1}{(1 + \omega^2)(4 + \omega^2)}$ and $\phi_{sn}(\omega) = 0$. A differential equation representation of this process is given by

$$\dot{x}(t) = \begin{bmatrix} -1 & 0 & 0 \\ 0 & -2 & 1 \\ 0 & 0 & -1 \end{bmatrix} x(t) + \begin{bmatrix} 1 & 0 \\ 0 & 0 \\ 0 & 1 \end{bmatrix} u(t)$$

$$y(t) = [1 \ \ 1 \ \ 0]x(t)$$

where $x(t)$ is a 3-vector; $u(t)$ is a 2-vector such that

$$\begin{aligned} Eu_1(t_1)u_1(t_2) &= \delta(t_2 - t_1) \\ Eu_2(t_1)u_2(t_2) &= \delta(t_2 - t_1) \\ Eu_1(t_1)u_2(t_2) &= 0 \end{aligned}$$

▶ and $y(t) = v(t)$; $\big(x_1(t) = s(t); \ x_2(t) = n(t)\big)$.

3.3 Difference Equation Model of Random Sequences

In the same manner as in the preceding section, discrete random sequences associated with practical situations can be modeled by the following vector difference equation

$$x(k + 1) = f[x(k), u(k), k]$$
$$y(k) = g[x(k), v(k), k] \tag{3.6}$$

where x and f are n-dimensional vectors; u and v are, respectively, r- and q-dimensional independent random sequences; g is an s-dimensional vector.

▶ example

The scalar signal $s(k)$ is a stationary random sequence with the properties

$$Es(k) = 0$$
$$Es(k)s(j) = e^{-2|j-k|}$$

A difference equation model of this process can easily be obtained.*

$$x(k + 1) = e^{-2}x(k) + \sqrt{1 - e^{-4}}\,u(k)$$
$$y(k) = x(k)$$
$$Eu(k)u(j) = \Delta(j - k)$$
$$Ex(0) = 0$$
$$Ex^2(0) = 1$$

where $\Delta(j - k)$ is the Kronecker delta, i.e.,

$$\Delta(j - k) = 1 \quad j = k$$
$$ = 0 \quad j \neq k$$

◀

3.4 Discussion of Differential and Difference Equation Models

A logical choice of a specific mathematical model of any physical process can only be made in accordance with a particular application involving the model. Yet certain models are to be preferred because significant properties of the process are characterized more explicitly or more naturally by these models. Accordingly, the following features of differential or difference equation models of random processes and sequences are cited.

1. In many practical cases the differential/difference equation model represents the operation of the actual physical system producing the random process. Consequently, any necessary approximations of the model for purposes of simplification in computational or analytical manipulations can be examined and justified by means of their implications in connection with the physical system.

---

* For example, $\mathscr{L}$-transformation can be used in a manner similar to usage of the Laplace transform in Example 1 of Section 3.2.

2. The variables $x(t)$ and $x(k)$ in the differential and difference equations models are vector Markov processes and sequences respectively. This fact is probably the basic motivation behind adoption of these models since it introduces significant simplifications in various mathematical operations.
3. This form of model can represent a variety of processes such as stationary gaussian, nonstationary gaussian, nonstationary non-gaussian, etc.
4. It is ideally suited for computer applications, e.g., if it is desired to generate the process by means of simulation.

3.5 Parameter Representation by the Model

In previous sections we have seen that many random processes, whether signal or noise, can be represented by the solution of a differential equation. In some cases the process is a function of one or more constant parameters. Let us denote these parameters by a constant vector γ. The representation (3.1) then assumes the following form

$$\dot{x}(t) = f[x(t), t, \gamma] + Bu(t)$$
$$y(t) = g[x(t), v(t), t] \tag{3.7}$$

We can represent γ as the solution of a number (equal to dimension of γ) of additional differential equations of the form

$$\dot{x}_{n+1} = 0$$
$$\dot{x}_{n+2} = 0 \tag{3.8}$$
$$\vdots$$

so that the solutions $x_{n+1}(t) = \gamma_1$, $x_{n+2}(t) = \gamma_2, \ldots$, etc., are all constants equal to the initial conditions γ. If these equations are added to (3.8), it assumes the form

$$\dot{x}(t) = f[x_1(t), \ldots, x_n(t), x_{n+1}(t), \ldots, t] + Bu(t)$$
$$y(t) = g[x(t), v(t), t] \tag{3.9}$$

where $x(t)$ is an augmented vector including the terms $x_{n+1}(t), \ldots$ in (3.8). Equation (3.9) now has the same form as (3.1) and consequently (3.1) can represent the system parameters as well.

▶ **example 1**

Let $n(t)$ be a random process with an unknown mean denoted by γ and where $n(t) - \gamma$ has the spectral density $\phi(\omega) = \dfrac{1}{1 + \omega^2}$. The following is a differential equation model for this process

$$\dot{x} = \begin{bmatrix} -1 & 1 \\ 0 & 0 \end{bmatrix} x(t) + \begin{bmatrix} 1 \\ 0 \end{bmatrix} u(t)$$

$$y(t) = \begin{bmatrix} 1 & 1 \end{bmatrix} x(t)$$

$$Eu(t_1)u(t_2) = \delta(t_2 - t_1)$$

▶ where $y(t) = n(t)$.

▶ **example 2**

Let $s(t)$ be the output of a linear constant-coefficient system with input $u(t)$, $Eu(t_1)u(t_2) = \delta(t_2 - t_1)$ and an unknown parameter γ as shown in Figure 3.3.

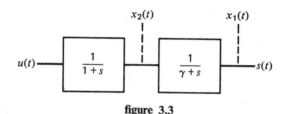

figure 3.3

The following set of differential equations can be easily derived

$$\dot{x}_1(t) = -x_3(t)x_1(t) + x_2(t)$$

$$\dot{x}_2(t) = -x_2(t) + u(t)$$

$$\dot{x}_3(t) = 0$$

$$y_1(t) = x_1(t)$$

where $y(t) = s(t)$. It is to be noted that, because of inclusion of the parameter γ, the resulting differential equations are (at least formally)
▶ nonlinear.

3.6 Linear Models

A special case of equation (3.1) can be written as

$$\dot{x}(t) = A(t)x(t) + B(t)u(t) \tag{3.10}$$

$$y(t) = C(t)x(t) + D(t)v(t) \tag{3.11}$$

where $A(t)$, $B(t)$, $C(t)$, and $D(t)$ are $n \times n, n \times r, s \times n$, and $s \times q$ matrices respectively. Equation (3.10) is a linear differential equation and has the solution for $t > 0$

$$x(t) = X(t)\left\{ x_0 + \int_0^t X^{-1}(s)B(s)u(s)\,ds \right\}; \qquad t \geq 0 \qquad (3.12)$$

where $X(t)$ is a nonsingular matrix called the transition matrix and satisfies the following matrix differential equation (see Section 1.17)

$$\dot{X}(t) = A(t)X(t)$$
$$X(0) = I \text{ identity matrix} \qquad (3.13)$$

A formal substitution of (3.12) and (3.13) into (3.10) can be used to verify that (3.12) is the solution to (3.10) with the initial condition

$$x(0) = x_0$$

For simplicity in the equations and without any loss of generality,* the initial time is assumed zero. For the case when $A(t)$ is a constant matrix $X(t)$ assumes the form

$$X(t) = e^{At} \stackrel{\Delta}{=} I + tA + \frac{t^2}{2!}A^2 + \cdots \qquad (3.14)$$

The value of the initial condition $x(0) = x_0$ is, in certain problems, independent of $u(t)$. (It may be a known or unknown constant vector or may be specified by a given probability density function.) On the other hand, if x_0 is caused by the application of some random function $u(t)$, $-\infty < t < 0$ in (3.10) then we have (assuming the initial condition at $t = -\infty$ is zero):

$$x(0) = x_0 = \int_{-\infty}^{0} X^{-1}(s)B(s)u(s)\,ds \qquad (3.15)$$

In this case the mean and covariance matrix of x_0 can be obtained from (3.15). Let

$$Eu(t) = 0$$
$$Eu(t_1)u'(t_1) = Q\delta(t_2 - t_1)$$
$$Q \text{ is an } r \times r \text{ matrix}$$

Hence

$$Ex_0 = \int_{-\infty}^{0} X^{-1}(s)B(s)Eu(s)\,ds = 0$$
$$Ex_0 x_0' = \int_{-\infty}^{0} X^{-1}(s)B(s)QB'(s)X^{-1'}(s)\,ds \qquad (3.16)$$

Since we have computed the mean and covariance matrix of x_0, the mean and

* It is assumed that the differential equation can be translated so as to have $t_0 = 0$.

covariance matrix of $x(t)$ can be obtained from (3.12)

$$Ex(t) = X(t)\left\{Ex_0 + \int_0^t X^{-1}(s)B(s)Eu(s)\,ds\right\} \qquad (3.17)$$

And when $Eu(t) = 0$

$$Ex(t)x'(t) = X(t)Ex_0x_0'X(t) + \int_0^t X(t)X^{-1}(s)B(s)QB'(s)X^{-1'}(s)X'(t)\,ds \qquad (3.18)$$

▶ **example 1**

Given

$$A(t) = -1$$
$$B(t) = 1$$
$$C(t) = 1$$
$$D(t) = 0$$
$$Q = 1$$
$$x(0) = 0$$

Let $u(t)$ be a stationary input $-\infty \le t \le +\infty$ having zero-mean with covariance matrix $Q\,\delta(t_2 - t_1)$. It is desired to derive the autocorrelation function of the output $y(t)$. Clearly $y(t)$ in (3.10, 3.11) is a zero-mean, but nonstationary random process (the process will become stationary as the transients die out). Using the given coefficients, we can compute

$$\varphi_y(t, \tau) = E[y(t)y(t + \tau)] = Ex(t)x(t + \tau)$$

hence

$$\varphi_y(t, \tau) = \int_0^t X(t)X^{-1}(s)BQB'X^{-1'}(s)X'(t + \tau)\,ds, \qquad \tau \ge 0$$

where

$$X(t) = e^{At} = e^{-t}$$

It follows that for $\tau \ge 0$

$$\varphi_y(t, \tau) = \tfrac{1}{2}e^{-\tau}[1 - e^{-2t}]$$

Let us denote $\lim_{t\to\infty} \varphi_y(t, \tau)$ by $\varphi_y(\tau)$. Therefore

$$\varphi_y(\tau) = \tfrac{1}{2}e^{-\tau} \qquad \tau \ge 0$$

Similarly

$$\varphi_y(\tau) = \tfrac{1}{2}e^{\tau} \qquad \tau \le 0$$

Then

▶
$$\varphi_y(\tau) = \tfrac{1}{2}e^{-|\tau|}$$

A linear model for a random sequence has the form

$$x(k + 1) = A(k)x(k) + B(k)u(k) \qquad (3.19)$$

$$y(k) = C(k)x(k) + D(k)v(k) \qquad (3.20)$$

where $A(k), B(k), \ldots$, etc., are defined the same as in (3.10) and (3.11) except here k is a discrete variable $(k = , \ldots, -1, 0, +1, \ldots)$. The solution to (3.19) can be written as a "convolution sum" which is the discrete version of (3.12):

$$x(1) = A(1)x(0) + B(0)u(0),$$

$$x(k) = \left(\prod_{i=k-1}^{0} A(i) \right) x(0) + B(k-1)u(k-1)$$

$$+ \sum_{i=0}^{k-2} \left(\prod_{j=k-1}^{i+1} A(j) \right) B(i)u(i), \qquad k \geq 2 \quad (3.21)$$

$$\prod_{i=k-1}^{0} A(i) \triangleq A(k-1)A(k-2) \cdots A(0)$$

where the matrix products are carried out with a decreasing index. The mean and covariance matrix of $x(k)$ in terms of those of $u(k)$ can be derived from (3.21) by essentially the same derivation that was used for continuous systems.

In a number of applications it is desired to represent a continuous process such as one defined by (3.10) by a discrete process (3.19) so that the solution of (3.19) be identical (in a statistical sense) with the solution of (3.10) at times $t = kT$ where T is a given constant referred to as a sampling period. One application is when it is desired to generate the process (3.10) by means of a digital computer. The problem is to determine $A(k)$, $B(k)$, and $Eu(k)u'(k)$ in terms of $A(t)$, $B(t)$, and $Eu(t)u'(t)$.

From (3.12)

$$x(kT) = X(kT)\left\{ x_0 + \int_0^{kT} X^{-1}(s)B(s)u(s)\, ds \right\} \quad (3.22)$$

$$x[(k+1)T] = X(k+1)T\left\{ x_0 + \int_0^{(k+1)T} X^{-1}(s)B(s)u(s)\, ds \right\} \quad (3.23)$$

Substituting (3.22) into (3.23)

$$x(k+1)T = [X(k+1)T \cdot X^{-1}(kT)]x(kT)$$

$$+ X(k+1)T \cdot \int_{kT}^{(k+1)T} X^{-1}(s)B(s)u(s)\, ds \quad (3.24)$$

Hence, if $u(s)$ has zero-mean, we can compute the mean of $x(k+1)T$ as

$$Ex(k+1)T = [X(k+1)TX^{-1}(kT)]Ex(kT) \quad (3.25)$$

and the covariance of $x(k+1)T$ as

$$E[x(k+1)T - Ex(k+1)T][x(k+1)T - Ex(k+1)T]'$$

$$= \int_{kT}^{(k+1)T} X(k+1)TX^{-1}(s)B(s)QB'(s)X^{-1'}(s)X'(k+1)T\, ds \quad (3.26)$$

By evaluating $Ex(k+1)$ and $E[x(k+1) - Ex(k+1)][x(k+1) - Ex(k+1)]'$ from (3.19) and equating with (3.25) and (3.26) respectively we get the desired results, namely,

$$A(k) = X(k+1)TX^{-1}(kT) \tag{3.27}$$

$EB(k)u(k)u'(k)B'(k)$

$$= \int_{kT}^{(k+1)T} X(k+1)TX^{-1}(s)B(s)QB'(s)X^{-1'}(s)X'(k+1)T \, ds \tag{3.28}$$

For the case where $A(t)$ and $B(t)$ are constant matrices and using (3.14) we have

$$A(k) = e^{AT} \tag{3.29}$$

$$EB(k)u(k)u'(k)B'(k) = \int_0^T e^{A(T-s)}BQB'e^{A'(T-s)} \, ds \tag{3.30}$$

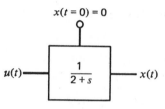

x(t = 0) = 0

$u(t)$ ——— $\dfrac{1}{2+s}$ ——— $x(t)$

figure 3.4

▶ **example 2**

A random process $x(t)$ is defined by the output of the system in Figure 3.4. The input $u(t)$ is a zero-mean white gaussian process whose covariance is

$$Eu(t_1)u(t_2) = \delta(t_2 - t_1)$$

The system is at rest for $t < 0$ and the input is applied at $t = 0$. It is desired to generate this process on a digital computer such that the generated sequence be (statistically) identical with the process $x(t)$ at $t = 0, 1, 2, \ldots$ seconds.

From (3.19), (3.29), and (3.30) the recursive relationship to be simulated on the computer is

$$x(k+1) = Ax(k) + u(k)$$

$$A = e^{-2}$$

▶
$$Eu(k)u'(k) = \frac{1 - e^{-4}}{4}$$

3.7 Observability

A question dealing with the representation (3.10) and (3.11) and closely related to the ability to estimate the value of $x(t)$ can now be asked. If $u(t)$,

$v(t)$, and $y(t)$ are known (or monitored) over an interval $0 \leq t \leq t_1$, what conditions should $A(t)$, $B(t)$, $C(t)$, and $D(t)$ satisfy in order to be able to solve for $x(t)$ during this interval with any arbitrary (unknown) initial condition $x(0)$. Clearly, if we are not able to derive a value for $x(t)$ when all the random inputs are assumed known $\big(u(t)$ and $v(t)\big)$ we surely cannot arrive at a reasonable estimate of $x(t)$ in the actual case when the observation is limited to $y(t)$ only.

Since $u(t)$ and $v(t)$ are known, their contribution can be eliminated from $y(t)$, and the above problem is restated as:

Given

$$\dot{x}(t) = A(t)x(t)$$
$$y(t) = C(t)x(t) \tag{3.31}$$
$$y(t); \quad 0 \leq t \leq t_1$$

Under what conditions can we solve for $x(t)$, $0 \leq t \leq t_1$? From (3.12)

$$x(t) = X(t)x(0) \tag{3.32}$$

Clearly, the knowledge of $x(t)$ for any t in the interval $\big($such as $x(t_1)\big)$ is sufficient to solve for $x(t)$, since from (3.32)

$$x(t_1) = X(t_1)X^{-1}(t)x(t) \tag{3.33}$$

In addition, substituting into (3.31) we get

$$y(t) = C(t)X(t)X^{-1}(t_1)x(t_1) \tag{3.34}$$

In order to be able to solve for all possible values of $x(t_1)$ $\big($or equivalently $x(0)\big)$ it is necessary that the columns of the matrix $C(t)X(t)X^{-1}(t_1)$ be such that there exist no nonzero $x(t_1)$ for which $y(t)$ will be identically zero over the entire interval $[0, t_1]$. Clearly in such a case we cannot solve for $x(t_1)$. This linear independence condition is expressed by the equivalent condition that the matrix M defined by

$$M = \int_0^{t_1} X^{-1'}(t_1)X'_{(t)}(t)C'(t)C(t)X(t)X^{-1}(t_1) \, dt \tag{3.35}$$

be positive definite.* Hence M^{-1} exists and from (3.34)

$$x(t_1) = M^{-1} \int_0^{t_1} X^{-1'}(t_1)X'(t)C'(t)y(t) \, dt \tag{3.36}$$

* For all arbitrary constant vectors α we have

$$\alpha'M\alpha = \int_0^{t_1} \| C(t)X(t)X^{-1}(t_1)\alpha \|^2 \, dt$$

Since the integrand cannot be identically zero over the entire interval $[0, t_1]$ unless $\alpha \equiv 0$, therefore, $\alpha'M\alpha > 0$ for $\alpha \neq 0$. The matrix M is positive definite by definition.

which proves that the condition of M being positive definite is also sufficient to be able to solve for $x(t_1)$. When every initial state $x(0)$ can be determined from $y(t)$, $0 \leq t \leq t_1$ or, equivalently, when M is positive definite, (3.31) is said to be "completely observable."

In the case of constant-coefficient systems (i.e., $A(t)$ and $C(t)$ in (3.31) are constant matrices) the condition that M be positive definite can be further simplified.

From the Cayley-Hamilton theorem (for matrices) every constant matrix satisfies its own characteristic equation, i.e., we can write

$$A^n + \alpha_1 A^{n-1} + \cdots + \alpha_n I = 0 \tag{3.37}$$

when $\alpha_1, \ldots, \alpha_n$ are constants which are the coefficients of corresponding powers of the scalar λ in the characteristic polynomial $|(A - \lambda I)|$. Post-multiplying (3.37) by $e^{AT}e^{-AT_1}x(t_1)$ and pre-multiplying by C and using the result

$$\frac{d^n}{dt^n}e^{At} = e^{At}A^n = A^n e^{At}$$

and hence

$$CA^n = \left[\frac{d^n}{dt^n}Ce^{At}\right]e^{-At}$$

we get

$$\left\{\frac{d^n}{dt^n}Ce^{+At}e^{-At_1} + \alpha_1 \frac{d^{n-1}}{dt^{n-1}}Ce^{+At}e^{-At_1} + \cdots\right\}x(t_1) = 0 \tag{3.38}$$

Now the preceding condition on columns of the matrix $C(t)X(t)X^{-1}(t_1) = Ce^{+At}e^{-At_1}$ leads to the condition that the solution $Ce^{At}e^{-At_1}x(t_1)$ of (3.38) not be identically zero. This in turn is satisfied if at least one of the terminal (or initial) conditions of the differential equation (3.38) is nonzero, i.e.,

$$\frac{d^k}{dt^k}Ce^{+A(t-t_1)}\bigg|_{t=t_1} x(t_1) \neq 0 \qquad \text{for some } k$$

or

$$CA^k x(t_1) \neq 0 \qquad \text{for some } k, \, 0 \leq k \leq n-1 \tag{3.39}$$

In a compact form (3.39) can be expressed as the condition that the rank of the following matrix be equal to n

$$N \triangleq [C', A'C', A'^2C', \ldots, A'^{n-1}C'] \tag{3.40}$$

To verify this statement recall that if the $n \times sn$ dimensional matrix N has rank n then

$$N'x(t_1) = x'(t_1)N \neq 0 \tag{3.41}$$

for all values of the n-dimensional vector $x(t_1) \neq 0$. Substituting for N yields

$$[x'(t_1)C', x'(t_1)A'C', \ldots, x'(t_1)A'^{n-1}C'] \neq 0 \tag{3.42}$$

Clearly (3.42) is equivalent to the condition prescribed by (3.39).

PROBLEMS

1. Derive a differential equation model for the random process $s(t)$; $t \geq 0$ where

$$Es(t) = e^{-t} \qquad t \geq 0$$

$$E[s(t_1) - e^{-t_1}][s(t_2) - e^{-t_2}] = e^{-|t_2 - t_1|}$$

2. A stationary discrete process $x(k)$ is obtained by sampling a continuous stationary process $x(t)$ at time intervals $t = k$, $k = 1, \ldots$ where

$$Ex(t) = 1$$

$$E[x(t_1) - 1][x(t_2) - 1] = e^{-|t_2 - t_1|}$$

Indicate how we can generate a sample function of $x(k)$ from a stationary zero-mean white noise sequence $u(k)$

$$Eu(k) = 0$$

$$Eu(k_1)u(k_2) = \Delta(k_2 - k_1)$$

by means of a digital computer.

3. A system has the transfer function

$$\frac{K(s + a)}{s(s + 1)}$$

where K and a are unknown constant parameters. The input to the system is unit step. Show that a linear differential equation model can be constructed if we allow functions of K and a to appear as initial conditions of the model. Discuss why a similar linear model cannot be constructed for Example 2 of Section 3.5 even if we replace the random input by unity.

4. Suggest an approach for approximately representing, by means of a discrete model, the nonlinear process

$$\dot{x}(t) = f[x(t), t] + Bu(t) \qquad x(0) = 0$$

where x and f are n-vectors and where u, an r-vector, is a white noise process. The form of the discrete model to be chosen is to be

$$x(k + 1) = g(x(k), k) + Cu(k)$$

Evaluate the pertinent parameters of the discrete model. Apply your procedure to a model of a linear system, and compare the results with those of Section 3.6.

5. Give two examples of systems being completely observable and not observable.

6. Verify the value of the covariance matrix given at the end of Example 1, Section 3.2.

SELECTED READINGS

Davenport, W. B., and W. L. Root, *An Introduction to the Theory of Random Signals and Noise*, McGraw-Hill, New York, 1958.

Kreindler, E., P. E. Sarachik, "On the Concepts of Controllability and Observability of Linear Systems," *IEEE Transactions on Automatic Control*, Vol. AC-9, 1964.

Lanning, J. H., Jr., and R. H. Battin, *Random Processes in Automatic Control*, McGraw-Hill, New York, 1956.

Papoulis, A., *Probability, Random Variables, and Stochastic Processes*, McGraw-Hill, New York, 1965.

chapter 4/ LINEAR ESTIMATORS

4.1 Introduction

The concept of optimal estimation was introduced in Chapter 2. This chapter is concerned with the derivation of a particular class of optimal estimators which yield a linear function of observation data as the estimate. As discussed before, in certain cases such as gaussian processes (or sequences) the best linear estimator is best of all estimators (linear or nonlinear) with respect to very general optimization criteria. There are other cases where we are merely looking for the best linear estimator either because of the simplicity in mechanization and construction which is generally associated with linear systems or just to limit the domain over which the search for the optimal estimator is to be carried out. Linear estimators for both discrete and continuous nonstationary processes are considered in this chapter. The observations of these processes are given either at a finite number of samples or over an interval of finite length. The case of stationary processes and infinite observation time is treated as a special case of the general results.

4.2 Discrete Estimation, Problem Statement

The discrete random process is defined by

$$x(k + 1) = Ax(k) + Bu(k) \qquad (4.1)$$

$$y(k) = Cx(k) + Dv(k) \qquad (4.2)$$

$$Eu(k) = Ev(k) = 0 \qquad (4.3)$$

$$Eu(k_1)u'(k_2) = K\,\Delta(k_2 - k_1) \qquad (4.4)$$

$$Ev(k_1)v'(k_2) = L\,\Delta(k_2 - k_1) \qquad (4.5)$$

where A, B, C, D, K, L are $n \times n$, $n \times r$, $s \times n$, $s \times q$, $r \times r$, and $q \times q$ matrices respectively. All these matrices are, in general, functions of k and the argument has been dropped here for simplicity in notation.* The term $\Delta(k_2 - k_1)$ is the Kronecker delta function, and $y(k)$ is defined as the observation. The initial value $x(0)$ is assumed a random vector with zero-mean and known covariance matrix $P(0)$.

It is desired to find the estimate of the n-vector $x(k + 1)$ denoted by $\hat{x}(k + 1)$ which is a linear function of observations $y(0), \ldots , y(k)$ minimizing

$$E[x(k + 1) - \hat{x}(k + 1)]'Q[x(k + 1) - \hat{x}(k + 1)] \qquad (4.6)$$

where Q is a symmetric positive definite matrix. Q in general could be a function of k. The cases where it is desired to obtain $\hat{x}(k + i)$, for $i > 1$ and $i = 0$ given $y(0), \ldots , y(k)$, can be handled in similar manner and will be discussed later in this chapter. Note that this estimation problem involves finding a "one-step predictor," since $\hat{x}(k + 1)$ is to be estimated from $y(0), \ldots , y(k)$. The expectation in (4.6) is with respect to u, v, and $x(0)$.

4.3 Principle of Orthogonality, Discrete Case

Since $\hat{x}(k + 1)$ was restricted to be a linear function of $y(0), \ldots , y(k)$ it can be written as

$$\hat{x}(k + 1) = \sum_{i=0}^{k} \alpha(i)y(i) \qquad (4.7)$$

where each $\alpha(i)$ is an $n \times s$ matrix and each $y(i)$ is an s-vector. The minimum of the quadratic function (4.6) is obtained by substituting (4.7) into (4.6) and differentiating with respect to each element of matrices $\alpha(i)$ $\big($namely, $\alpha_{jl}(i)$; $j = 1, \ldots , n, l = 1, \ldots , s; i = 0, \ldots , k\big)$. This yields a total of $ns(k + 1)$ equations, which, in compact form are given by†

$$E(Q + Q')[x(k + 1) - \hat{x}(k + 1)]y'(i) = 0, \qquad i = 0, \ldots , k$$

Since Q is positive definite it follows that

$$E[x(k + 1) - \hat{x}(k + 1)]y'(i) = 0, \qquad i = 0, \ldots , k \qquad (4.8)$$

In deriving (4.8) it was assumed that the order of expectation and the differentiation with respect to the elements of $\alpha(i)$ can be interchanged, since these elements are not random variables. Equation (4.8) is the "orthogonality principle," which states that the linear estimate $\hat{x}(k + 1)$ which minimizes the

---

* Also for simplicity in notation $u(k)$ and $v(k)$ are assumed uncorrelated. The procedure is almost identical when this assumption is eliminated.

† Alternatively we can arrive at (4.8) by substituting $\hat{\alpha}(i) + \varepsilon\alpha(i)$ for $\alpha(i)$ and setting the derivative of (4.6) with respect to ε zero (see Section 1.40).

quadratic cost given by (4.6) is such that the estimation error $[x(k+1) - \hat{x}(k+1)]$ is uncorrelated with every one of the observations $y(0), \ldots, y(k)$. Furthermore, it can be shown that if a linear estimator (defined by a set of values for $\alpha(i)$ in (4.7)) satisfies (4.8) it then minimizes the quadratic cost given by (4.6). This statement means, mathematically, that (4.8) is a sufficient as well as necessary condition for minimization of (4.6). The orthogonality condition given by (4.8) is also a necessary condition for achieving the minimum of the quadratic function

$$z'E[x(k+1) - \hat{x}(k+1)][x(k+1) - \hat{x}(k+1)]'z$$

for any arbitrary nonzero vector z and where $\hat{x}(k+1)$ is given by (4.7). To verify this let us assume $\hat{\alpha}(i)$ yields the minimum. Consequently,

$$z'E\left\{x(k+1) - \sum_0^k \hat{\alpha}(i)y(i)\right\}\left\{x(k+1) - \sum_0^k \hat{\alpha}(i)y(i)\right\}z'$$

$$\leq z'E\left\{x(k+1) - \sum_0^k [\hat{\alpha}(i) + \varepsilon(i)\alpha(i)]y(i)\right\}$$

$$\times \left\{x(k+1) - \sum_0^k [\hat{\alpha}(i) + \varepsilon(i)\alpha(i)]y(i)\right\}'z$$

for any arbitrary $\alpha(i)$ and scalar values $\varepsilon(i)$. Expanding the right hand side of this inequality and considering $\varepsilon(i)$'s to be arbitrarily small (so that terms in $\varepsilon^2(i)$ drops out) yields

$$z'E\{x(k+1) - \hat{x}(k+1)\} \sum_1^k y'(j)\alpha'(j)\varepsilon(j)z \geq 0$$

This inequality is to be satisfied for any set of values $\varepsilon(i)$ such as

$$\varepsilon(j) = 0 \quad j \neq i$$
$$\varepsilon(j) \neq 0 \quad j = i$$

Hence a necessary condition for the minimum of the quadratic function is

$$z'E\{x(k+1) - \hat{x}(k+1)\}y'(i)\alpha'(i)\varepsilon(i)z \geq 0 \quad i = 0, \ldots, k$$

Since $\varepsilon(i)$ can be either positive or negative, this inequality can only be satisfied as equality

$$z'E\{x(k+1) - \hat{x}(k+1)\}y'(i)\alpha'(i)z = 0 \quad i = 0, \ldots, k$$

Clearly the orthogonality condition given by (4.8) will satisfy this relationship for all $\alpha(i)$ and z as required.

As an example, let us choose z to have all zero components except z_j and set

$$z_j = 1$$

It follows that

$$z'E\{x(k+1) - \hat{x}(k+1)\}\{x(k+1) - \hat{x}(k+1)\}'z$$
$$= E\{x_j(k+1) - \hat{x}_j(k+1)\}^2$$

Therefore (4.8) is the condition necessary for achieving a minimum variance estimate of any one component of $x(k+1)$. It follows, further, that (4.8) is a necessary condition for achieving minimum variance estimate for *all* components of $x(k+1)$ *simultaneously*.

4.4 Discrete Estimator (Kalman)

We shall now solve the problem introduced in Section 4.2. Since the estimator was required to be linear, let us conjecture that it have the linear form

$$\hat{x}(k+1) = F_1\hat{x}(k) + F_2y(k) \tag{4.9}$$

where the matrices F_1 and F_2 may be functions of k, but are not random [they are, however, determined by the statistics associated with (4.1)–(4.5), which are assumed known]. In order to verify this conjecture it is sufficient to show that we can find matrices F_1 and F_2 so that the corresponding $\hat{x}(k+1)$ satisfies (4.8). The reason for this choice is the recursive nature of (4.9). If we could obtain the estimate $\hat{x}(k+1)$ in the recursive form (4.9), it would not be necessary to process the entire observation data $y(0), \ldots, y(k)$; rather, the previous estimate $\hat{x}(k)$ and the last observation $y(k)$ would suffice.

Equation (4.8) can be written as

$$E[x(k+1) - \hat{x}(k+1)]y'(i) = 0 \qquad i = 0, \ldots, k-1 \tag{4.10}$$

$$E[x(k+1) - \hat{x}(k+1)]y'(k) = 0 \tag{4.11}$$

Let us substitute for $x(k+1)$ from (4.1) and $\hat{x}(k+1)$ from (4.9) into (4.10). We note from (4.1) and (4.2) that the data $y(0), \ldots, y(k)$ do not involve $u(k)$. Since the $u(i)$ and $v(i)$ sequences are uncorrelated, it follows that $Eu(k)y'(i) = 0$, $0 \le i \le k$. Using this fact, we obtain

$$E[Ax(k) - F_1\hat{x}(k) - F_2y(k)]y'(i) = 0, \qquad i = 0, \ldots, k-1 \tag{4.12}$$

Equation (4.8) also holds for the optimal estimate at the previous step. Hence

$$E[x(k) - \hat{x}(k)]y'(i) = 0 \qquad i = 0, \ldots, k-1 \tag{4.13}$$

Substituting for $y(k)$ from (4.2) into (4.12) yields

$$E[Ax(k) - F_1\hat{x}(k) - F_2Cx(k) - F_2Dv(k)]y'(i) = 0$$
$$i = 0, \ldots, k-1 \tag{4.14}$$

We note from (4.1) and (4.2) that the data $y(0), \ldots, y(k-1)$ do not involve $u(k)$. Again, since the $u(i)$ and $v(i)$ sequences are uncorrelated, it follows that $Ev(k)y'(i) = 0$, $i = 0, \ldots, k-1$. Using this fact and substituting for $Ex(k)y'(i)$ from (4.13) into (4.14), we obtain

$$E[A - F_1 - F_2C]\hat{x}(k)y'(i) = [A - F_1 - F_2C]E\hat{x}(k)y'(i) = 0$$
$$i = 0, \ldots, k-1$$

Irrespective of what values $E\hat{x}(k)y'(i)$ assume, this equation can be satisfied if the matrix F_1 is chosen to be

$$F_1 = A - F_2C \qquad (4.15)$$

Clearly this choice for F_1 causes (4.9) to satisfy a portion of the condition given by (4.8), namely (4.10). F_2 now will be chosen such that (4.11) is satisfied.

Let us define

$$\hat{y}(k) \triangleq C\hat{x}(k)$$
$$\tilde{x}(k) \triangleq x(k) - \hat{x}(k) \qquad (4.16)$$
$$\tilde{y}(k) \triangleq y(k) - \hat{y}(k) = y(k) - C\hat{x}(k) \qquad (4.17)$$

The vector $\tilde{x}(k)$ is then the estimation error. Since $\tilde{y}(k)$ is linear in $\hat{x}(k)$, and because $\hat{x}(k)$ is by definition linear in $y(i)$, $0 \leq i \leq k-1$ from (4.8), we have necessarily that

$$E[x(k+1) - \hat{x}(k+1)]\hat{y}'(k) = 0 \qquad (4.18)$$

Subtracting (4.18) from (4.11) yields, by definition (4.17)

$$E[x(k+1) - \hat{x}(k+1)]\tilde{y}'(k) = 0 \qquad (4.19)$$

Substituting for $x(k+1)$ from (4.1) and $\hat{x}(k+1)$ from (4.9) and utilizing the definitions (4.17) it follows that

$$E[Ax(k) + Bu(k) - F_1\hat{x}(k) - F_2y(k)][y(k) - C\hat{x}(k)]' = 0$$

However, by the system structure we have

$$Eu(k)y'(k) = Eu(k)\hat{x}'(k) = 0$$

Thus we are left with

$$E[Ax(k) - F_1\hat{x}(k) - F_2y(k)][y(k) - C\hat{x}(k)]' = 0$$

We now substitute for $y(k)$ from (4.2) and F_1 from (4.15) and use definition (4.16) to get

$$E[(A - F_2C)\tilde{x}(k) - F_2Dv(k)][C\tilde{x}(k) + Dv(k)]' = 0 \qquad (4.20)$$

Let us define the following estimation error covariance matrix

$$P(k) \triangleq E\tilde{x}(k)\tilde{x}'(k) \qquad (4.21)$$

Note that $P(k)$ is the covariance matrix of the error incurred by using $\hat{x}(k)$ computed from data $y(0), \ldots, y(k-1)$ to estimate $x(k)$.

Now the expectation indicated in (4.20) involves $E\tilde{x}(k)v'(k)$ which is by definition (4.16) equal to $E[x(k) - \hat{x}(k)]v'(k)$. By the system structure (4.1) and (4.2), $x(k)$ is always uncorrelated with the $v(i)$ sequence, and $\hat{x}(k)$ does not involve $v(k)$, so it follows that $E\tilde{x}(k)v'(k) = 0$. Using this fact, together with the definition (4.21), the expectation in (4.20) yields $(A - F_2C)P(k)C' - F_2DLD' = 0$. This can be solved for F_2 yielding*

$$F_2(k) = AP(k)C'[CP(k)C' + DLD']^{-1} \qquad (4.22)$$

Here $F_2(k)$ is given in terms of $P(k)$. The covariance matrix $P(k)$ will now be shown to satisfy a difference equation which depends only on the system structure (4.1)–(4.5). We first use the definitions (4.16) and (4.21) to define $P(k + 1)$, and substitute from (4.1) and (4.9) to obtain

$$P(k + 1) = E[Ax(k) + Bu(k) - F_1\hat{x}(k) - F_2y(k)]$$
$$\times \ [Ax(k) + Bu(k) - F_1\hat{x}(k) - F_2y(k)]'$$

After substituting for F_1 from (4.15) and $y(k)$ from (4.2) and using the fact that random variables $\tilde{x}(k)$, $u(k)$, and $v(k)$ are mutually uncorrelated it follows that

$$P(k + 1) = [A - F_2(k)C]P(k)[A - F_2(k)C]'$$
$$+ BKB' + F_2(k)DLD'F_2'(k) \quad (4.23)$$

This completes the derivation of the discrete (Kalman) one-step predictor, which indeed has been implemented in the recursive form conjectured in (4.9). The results are the three equations (4.9), (4.22), and (4.23) repeated below, using the notational simplifications $F \triangleq F_2$

$$\hat{x}(k + 1) = [A - F(k)C]\hat{x}(k) + F(k)y(k) \qquad (4.24)$$

$$F(k) = AP(k)C'[CP(k)C' + DLD']^{-1} \qquad (4.25)$$

$$P(k + 1) = [A - F(k)C]P(k)[A - F(k)C]'$$
$$+ BKB' + F(k)DLD'F'(k) \quad (4.26)$$

The data sequence is indexed as $y(k)$, $0 \leq k \leq \infty$. The estimator of course gives $\hat{x}(k + 1)$, $0 \leq k \leq \infty$. The initial conditions $\hat{x}(0)$ and $P(0)$ should be specified as a part of available data. However, under very general conditions,

* When the observation $y(i)$ is vector-valued, the term in brackets in (4.22) becomes an $s \times s$ matrix which is assumed to be nonsingular. This assumption is certainly satisfied when $P(k)$ is positive definite and in addition all rows of C are linearly independent. The latter assumption implies that the observations (components of $y(i)$ vector) are not redundant. Notice that L is a covariance matrix and consequently is a non-negative matrix, so that DLD' is also non-negative. In the degenerate case when $CP(k)C'$ is singular the assumption that the bracket in (4.22) is nonsingular is still satisfied if DLD' is nonsingular.

it can be shown that (4.24), (4.25), and (4.26) represent an asymptotically stable system, and consequently the estimate $\hat{x}(k)$ becomes independent of the initial information $\hat{x}(0)$, $P(0)$ as k is increased.

It is important to notice that the equations for $F(k)$ and $P(k)$ are independent of the observations. Consequently $F(k)$ can be determined before any observations are made available.

▶ **example**

Let $y(k)$ be a scalar random sequence with unknown mean denoted by x and covariance function given by

$$E[y(k_1) - x][y(k_2) - x] = L\,\Delta(k_2 - k_1)$$

It is desired to estimate x (i.e., to predict x at $k + 1$) from the observation $y(0), \ldots, y(k)$. The model for the system can be given by

$$x(k + 1) = x(k)$$
$$y(k) = x(k) + v(k)$$
$$Ev(k_1)v(k_2) = L\,\Delta(k_2 - k_1)$$

Equation (4.25) then becomes

$$F(k) = \frac{P(k)}{P(k) + L}$$

and (4.26) then yields

$$P(k + 1) = \frac{LP(k)}{L + P(k)}$$

The preceding two equations along with (4.24) and $P(0)$ constitute the solution to the estimation problem. However, in this simple example further simplification is possible which reveals insight into the mechanism of the estimators. The nonlinear difference equation for $P(k)$ can be solved explicitly for $P(k)$ as a function of k, yielding

$$P(k) = \frac{L}{k + L/P(0)}$$

Consequently, (4.25) and (4.24) become

$$F(k) = \frac{1}{1 + \left(k + \dfrac{L}{P(0)}\right)}$$

$$\hat{x}(k + 1) = \frac{\left(k + \dfrac{L}{P(0)}\right)}{1 + \left(k + \dfrac{L}{P(0)}\right)}\,\hat{x}(k) + \frac{1}{1 + \left(k + \dfrac{L}{P(0)}\right)}\,y(k)$$

It may be verified, by direct substitution into the above relation, that $\hat{x}(k + 1)$ is given as an explicit function of the data $y(0), \ldots, y(k)$:

$$\hat{x}(k + 1) = \frac{1}{k + 1 + \dfrac{L}{P(0)}} \sum_{i=0}^{k} y(i), \qquad k \geq 0$$

This explicit formula for $x(k + 1)$ is consistent with the recursive formula given above if we take $\hat{x}(0) = 0$.

It is seen that as k approaches infinity $\hat{x}(k + 1)$ will approach the sample mean. Furthermore, $\hat{x}(k + 1)$ approaches the sample mean faster for larger values of $P(0)$ and smaller L, i.e., when we know that the initial estimate $\hat{x}(0) = 0$ is not good and the deviation of $y(t)$ about its mean has

▶ small covariance function. The sample mean is defined by $\lim\limits_{k \to \infty} \dfrac{1}{k} \sum\limits_{i=0}^{k} y(i)$.

4.5 Continuous Estimation, Problem Statement

Let the continuous random process $x(t)$ and the observation $y(t)$ be given by

$$\dot{x}(t) = Ax(t) + Bu(t) \tag{4.27}$$

$$y(t) = Cx(t) + Dv(t) \tag{4.28}$$

$$Eu(t) = Ev(t) = 0$$

$$Eu(t_1)u'(t_2) = K\,\delta(t_2 - t_1) \tag{4.29}$$

$$Ev(t_1)v'(t_2) = L\,\delta(t_2 - t_1) \tag{4.30}$$

$$Eu(t_1)v'(t_2) = M\,\delta(t_2 - t_1) \tag{4.31}$$

where A, B, C, D, K, L, and M are $n \times n, n \times r, s \times n, s \times q, r \times r, q \times q$, and $r \times q$ matrices, respectively. All these matrices in general are functions of t. The term $\delta(t_2 - t_1)$ is the Dirac delta function. The matrices K and L are covariance matrices and hence at least positive semi-definite,* but A, B, C, and D may be general matrices of appropriate dimension.

It is desired to find the estimate of the n-vector $x(t)$ denoted by $x(t)$ as a linear function (or, more precisely, a linear functional) of the observation $y(\tau)$, $0 \leq \tau \leq t$ which minimizes

$$E[x(t) - \hat{x}(t)]'Q[x(t) - \hat{x}(t)] \tag{4.32}$$

where Q is a positive definite (and, of course, nonsingular) matrix. Again Q in general is a function of t.

---

* In addition, to solve the problem posed in this section it will become necessary to assume DLD' to be nonsingular. This assumption will later be removed in Chapter 6.

The optimal continuous estimator can be obtained formally by applying a limiting process to the discrete optimal filter given by (4.25)–(4.27). However, mathematical difficulties will arise since the limit of a white noise sequence as sampling is increased does not approach a white noise process. In order to avoid this difficulty a direct derivation of the continuous estimator is given here. The reader may recognize a very close similarity between the derivations of continuous and discrete estimators.

4.6 Principle of Orthogonality, Continuous Case

The estimate $\hat{x}(t)$ was restricted to be a linear function of $y(\tau)$, $0 \leq \tau \leq t$. Consequently it can be written

$$\hat{x}(t) = \int_0^t \alpha(\tau) y(\tau) \, d\tau \tag{4.33}$$

where $\alpha(\tau)$ is an $n \times s$ matrix function in which each element is a function of τ. It is desired to find that $\alpha(\tau)$, denoted by $\alpha^0(\tau)$, which will minimize the quadratic function given by (4.32).

Using the ideas of calculus of variations (Section 1.42) let $\alpha^{00}(\tau)$ be any arbitrary function of τ and ε be an arbitrarily small scalar. Letting

$$\alpha(\tau) = \alpha^0(\tau) + \varepsilon \alpha^{00}(\tau)$$

and substituting this $\alpha(\tau)$ in the quadratic function (4.32) yields

$$E\left[x(t) - \int_0^t [\alpha^0(\tau) + \varepsilon\alpha^{00}(\tau)]y(\tau)\, d\tau \right]' Q\left[x(t) - \int_0^t [\alpha^0(\tau) + \varepsilon\alpha^{00}(\tau)]y(\tau)\, d\tau \right] \tag{4.34}$$

If $\alpha^0(\tau)$ yields the minimum value for (4.32), then the coefficient of the term in ε in the expansion of (4.34) must be zero, since ε can be chosen small with an arbitrary sign. It follows that

$$E\int_0^t y'(\tau)\alpha^{00'}(\tau)Q[x(t) - \hat{x}(t)]\, d\tau = 0 \tag{4.35}$$

Since $\alpha^{00}(\tau)$ is any arbitrary matrix function, the integrand of (4.35) should be zero for all τ; $0 \leq \tau \leq t$.

$$Ey'(\tau)\alpha^{00'}(\tau)Q[x(t) - \hat{x}(t)] = 0 \qquad 0 \leq \tau \leq t$$

Since the matrix $\alpha^{00}(\tau)$ is arbitrary, Q can be set equal to the matrix identity without loss of generality as long as Q is nonsingular. This yields

$$Ey'(\tau)\alpha^{00'}(\tau)[x(t) - \hat{x}(t)] = 0$$

or equivalently

$$E\sum_j \sum_i \alpha_{ij}^{00}(\tau)[x_i(t) - \hat{x}_i(t)]y_j(\tau) = 0 \qquad 0 \leq \tau \leq t$$

where the α_{ij}^{00} are the elements of the matrix α^{00}. Since this equation is to be satisfied for any function $\alpha_{ij}^{00}(\tau)$ we necessarily must have

$$E[x_i(t) - \hat{x}_i(t)]y_j(\tau) = 0 \qquad \text{all } i, j \text{ and } 0 \leq \tau \leq t$$

or compactly

$$E[x(t) - \hat{x}(t)]y'(\tau) = 0 \qquad 0 \leq \tau \leq t \qquad (4.36)$$

Equation (4.36) is the orthogonality principle in the continuous case and corresponds to (4.8) derived for discrete processes. It again states that the estimation error $[x(t) - \hat{x}(t)]$ incurred by using the optimal estimate $\hat{x}(t)$ is uncorrelated with the observation $y(\tau)$ for any specific value of τ in the interval of observation. Since (4.32) is quadratic it can be shown that (4.36) is the necessary and sufficient condition for achieving its minimum value.

4.7 Continuous Optimal Estimator (Kalman-Bucy)

The estimator will be conjectured to have the recursive form given by

$$\dot{\hat{x}}(t) = F_1\hat{x}(t) + F_2 y(t) \qquad (4.37)$$

In order to verify this statement it is sufficient to determine values for the matrix functions F_1 and F_2 such that (4.36) is satisfied. Using a procedure similar to that used for the discrete estimator, we first attempt to satisfy the orthogonality condition for $0 \leq \tau < t$ [analogous to (4.10)] and then consider the condition at $\tau = t$ [analogous to (4.1)].

Let us differentiate (4.36) with respect to t. For $\tau < t$ it follows that

$$E[\dot{x}(t) - \dot{\hat{x}}(t)]y'(\tau) = 0 \qquad 0 \leq \tau < t \qquad (4.38)$$

Substituting for $\dot{x}$ and $\dot{\hat{x}}(t)$ from (4.27) and (4.37) yields

$$E[Ax(t) + Bu(t) - F_1\hat{x}(t) - F_2 y(t)]y'(\tau) = 0 \qquad 0 \leq \tau < t \quad (4.39)$$

From (4.36) and (4.39), substituting for $y(t)$ from (4.28), and using the fact that

$$Eu(t)y'(\tau) = 0 \qquad \tau < t$$
$$Ev(t)y'(\tau) = 0 \qquad \tau < t$$

we obtain that (4.39) and hence (4.38) is satisfied if

$$F_1 = A - F_2 C \qquad (4.40)$$

To satisfy the orthogonality condition at $\tau = t$, we return to (4.36) and substitute for $y(\tau)$ from (4.28) and let $\tilde{x}(t) = x(t) - \hat{x}(t)$. It follows that

$$E\tilde{x}(t)[x'(\tau)C' + v'(\tau)D'] = 0 \qquad 0 \leq \tau \leq t \qquad (4.41)$$

Let us denote the transition matrix associated with (4.27) and (4.37) by $X(t)$ and $\hat{X}(t)$ respectively. The solution to (4.27) is then

$$x(t) = X(t)x(0) + \int_0^t X(t)X^{-1}(s)Bu(s)\,ds$$

and the solution to (4.37) for $\hat{x}(0) = 0$ is given by*

$$\hat{x}(t) = \int_0^t X(t)\hat{X}^{-1}(\rho)F_2(\rho)y(\rho)\,d\rho$$

The term $E\tilde{x}(t)v'(\tau)$ in (4.41) can now be evaluated

$$E\hat{x}(t)v'(\tau) = E[x(t) - \hat{x}(t)]v'(\tau) = EX(t)x(0)v'(\tau)$$

$$+ \int_0^t X(t)X^{-1}(s)BEu(s)v'(\tau)\,ds - \int_0^t \hat{X}(t)\hat{X}^{-1}(\rho)F_2(\rho)Ey(\rho)v'(\tau)\,d\rho \quad (4.42)$$

The first term in the right hand side of (4.42) is zero since $x(0)$ and $\mathrm{v}(\tau)$ are assumed uncorrelated. The second term from (4.31) becomes

$$\int_0^t X(t)X^{-1}(s)BEu(s)v'(\tau)\,ds = X(t)X^{-1}(\tau)BM \quad (4.43)$$

Substituting for $y(\rho)$ from (4.28) into the third term in the right hand side of (4.42) and considering that $Ex(\rho)v'(\tau) = 0$, $\rho < \tau$ it follows that

$$\int_0^t \hat{X}(t)\hat{X}^{-1}(\rho)F_2(\rho)Ey(\rho)v'(\tau)\,d\rho$$

$$= \hat{X}(t)\hat{X}^{-1}(\tau)F_2(\tau)DL + \int_\tau^t \hat{X}(t)\hat{X}^{-1}(\rho)F_2(\rho)CEx(\rho)v'(\tau)\,d\rho \quad (4.44)$$

The terms in the right hand side of (4.43) and (4.44) are continuous. Therefore, as τ approaches t in the limit, (4.43) becomes BM and (4.44) yields $F_2(t)DL$. Substituting these limits into (4.43) yields

$$E\tilde{x}(t)v'(t) = BM + F_2DL$$

Let us substitute this into (4.41) and replace $x'(t)$ in (4.41) by $\tilde{x}'(t) - \hat{x}'(t)$. Considering that $\hat{x}(t)$ is a linear function of $y(\tau)$, $0 \le \tau < t$ so that

$$E\tilde{x}(t)\hat{x}(t) = 0$$

it follows that

$$E\tilde{x}(t)\tilde{x}'(t)C' = -[BMD' - F_2(t)DLD'] \quad (4.45)$$

* The initial condition $\hat{x}(0) = 0$ is chosen when the *a priori* density $p[x(0)]$ has zero-mean. However, in general it is not necessary to assume $\hat{x}(0) = 0$. This point will be discussed later in this section.

Let us introduce the error covariance matrix $P(t)$ by

$$P(t) = E\tilde{x}(t)\tilde{x}'(t) \qquad (4.46)$$

Equation (4.45) can now be solved for F_2.

$$F_2(t) = [P(t)C' + BMD'][DLD']^{-1} \qquad (4.47)$$

The covariance matrix $P(t)$ will now be shown to satisfy a differential equation which depends only on the system structure. From (4.27), (4.28), and (4.37), and by substituting for F_1 we obtain a differential equation for the estimation error $\tilde{x}(t)$

$$\dot{\tilde{x}}(t) = [A - F_2(t)C]\tilde{x}(t) - F_2(t)Dv(t) + Bu(t)$$

This equation has a transition matrix identical with that of (4.37), namely $\hat{X}(t)$. Hence, solving for $\tilde{x}(t)$ yields

$$\tilde{x}(t) = \hat{X}(t)\tilde{x}(0) + \int_0^t \hat{X}(t)\hat{X}^{-1}(s)[-F_2(s)Dv(s) + Bu(s)]\,ds$$

Let us substitute this expression for $\tilde{x}(t)$ into (4.46). It follows that, since $\tilde{x}(0) = x(0) - \hat{x}(0)$ is assumed uncorrelated with $u(s)$ and $v(s)$,

$$P(t) = \hat{X}(t)P(0)\hat{X}'(t) + \int_0^t \hat{X}(t)\hat{X}^{-1}(s)[F_2(s)DLD'F_2'(s)$$
$$-F_2(s)DM'B' - BMD'F_2'(s) + BKB']\hat{X}^{-1'}(s)\hat{X}'(t)\,ds \qquad (4.48)$$

Differentiating both sides of (4.48), substituting for $\dot{\hat{X}}(t) = [A - F_2(t)C]\hat{X}(t)$, and reutilizing (4.48) yields

$$\dot{P}(t) = [A - F_2(t)C]P(t) + P(t)[A' - C'F_2'(t)] + F_2(t)DLD'F_2'(t)$$
$$- F_2(t)DM'B' - BMD'F_2'(t) + BKB' \qquad (4.49)$$

Substituting for $F_2(t)$ from (4.47) yields

$$\dot{P}(t) = AP(t) + P(t)A' - [P(t)C' + BMD']$$
$$\times [DLD']^{-1}[CP(t) + DM'B'] + BKB' \qquad (4.50)$$

This completes the derivation of the continuous estimator.* Substituting for the values of F_1 and F_2 in (4.37) yields

$$\dot{\hat{x}}(t) = [A - (P(t)C' + BMD')(DLD')^{-1}C]\hat{x}(t)$$
$$+ (P(t)C' + BMD')(DLD')^{-1}y(t) \qquad (4.51)$$

Equations (4.50) and (4.51) along with initial conditions $P(0)$ and $\hat{x}(0) = 0$ describe the operation of the estimator. The quantity $P(0)$, which is the *a*

* It is assumed that (4.50) has a unique solution.

priori covariance of $\bar{x}(0)$, is to be given as part of available data. It can easily be shown that if an *a priori* estimate is available for $\hat{x}(0)$ such as $\hat{x}^*(0)$ then it should be used as initial condition for (4.51) while $P(0)$ should be chosen as Cov $[x(t) - \hat{x}^*(0)]$.

▶ **example**

The following simple example is given to illustrate the application of results given by (4.50) and (4.51). Let $y(t)$ be a scalar white noise process with unknown mean x and given covariance function

$$Ey(t) = x$$
$$E[y(t_1) - x][y(t_2) - x] = L\,\delta(t_2 - t_1)$$

It is desired to estimate the mean x by processing the data $y(\tau)$, $0 \le \tau \le t$. A model for the system can be given by

$$\dot{x}(t) = 0$$
$$y(t) = x(t) + v(t)$$
$$Ev(t_2)v(t_1) = L\,\delta(t_2 - t_1)$$

Equation (4.50) will then become

$$\dot{P}(t) = -L^{-1}P^2(t)$$

or

$$\frac{dP(t)}{P^2(t)} = -L^{-1}\,dt$$

Integrating both sides yields

$$-P^{-1}(t) = -L^{-1}t + \text{Constant}$$

or, specifically,

$$P(t) = \frac{1}{L^{-1}t + 1/P(0)}$$

Substituting into (4.51) results in

$$\dot{\hat{x}}(t) = \frac{-1}{t + L/P(0)}\,\hat{x}(t) + \frac{1}{t + L/P(0)}\,y(t)$$
$$\hat{x}(0) = 0$$

This is a time-variable differential equation describing the estimator. Being a scalar equation, it can be solved for $\hat{x}(t)$. The transition matrix is easily obtained as

$$X(t) = \frac{L/P(0)}{t + L/P(0)}$$

Hence

$$\hat{x}(t) = \int_0^t \frac{L/P(0)}{t + L/P(0)} \cdot \frac{\tau + L/P(0)}{L/P(0)} \cdot \frac{1}{\tau + L/P(0)} \, y(\tau) \, d\tau$$

or

$$\hat{x}(t) = \frac{1}{t + L/P(0)} \int_0^t y(\tau) \, d\tau$$

It is seen that

$$\lim_{t \to \infty} \hat{x}(t) = \lim_{t \to \infty} \frac{1}{t} \int_0^t y(\tau) \, d\tau$$

which is a clearly expected result. Furthermore, it is seen that as the uncertainty about the initial estimate of the mean is decreased (i.e., $P(0)$ approaching zero) or as L is increased, the estimate $\hat{x}(t)$ remains insensitive to observations for a while but eventually approaches the limit ▶ shown above.

A number of properties of the estimator are worthy of further discussion. First it is necessary that the matrix DLD' be nonsingular. This implies that each element of the observation vector should contain a white noise term. This restriction will be later removed in Chapter 6. Furthermore, it should be noted that the estimator described by (4.51) is in general a time-varying linear system even when the random processes $u(t)$, $v(t)$ are stationary and matrices A, B, C, D, and Q are constants. This is because $y(t)$ was observed over a finite length of time $0 \le \tau \le t$. The case when the observation is given over an infinite interval will be discussed in Chapter 5. Finally, no restriction was put on the matrices A and B other than what is needed to arrive at a solution to linear differential equations. For example, the differential equation can be actually unstable and consequently the covariance matrix of $\hat{x}(t)$ can be unbounded, yet equations (4.50) and (4.51) are valid for any finite time t. Clearly in this case the $\lim_{t \to \infty} \hat{x}(t)$ does not have any meaning. The equation given by (4.50) is a matrix form of a Riccati differential equation. It is a very particular nonlinear differential equation and it will be discussed in the next section.

4.8 The Error Covariance Equation

The continuous estimation of the preceding section requires the solution of the matrix Riccati equation given by (4.50). It can be shown that this equation has a unique solution if $P(0)$ is positive semi-definite. Here it will be shown that a solution of the form

$$P(t) = X(t)Z^{-1}(t) \tag{4.52}$$

satisfies (4.50), where $X(t)$ and $Z(t)$ are given as solutions of linear matrix differential equations.

From (4.52)

$$\dot{X}(t) = \dot{P}(t)Z(t) + P(t)\dot{Z}(t) \qquad (4.53)$$

Substituting for $\dot{P}(t)$ from (4.51) and rearranging terms yields

$$\dot{X}(t) = A_x X(t) + B_x Z(t) + P[\dot{Z}(t) - A_z X(t) - B_z(t)] \qquad (4.54)$$

where

$$A_x = A - BMD'[DLD']^{-1}C \qquad (4.55)$$

$$B_x = BKB' - BMD'(DLD')^{-1}DM'B' \qquad (4.56)$$

$$A_z = C'[DLD']^{-1}C \qquad (4.57)$$

$$B_z = -A' + C'[DLD']^{-1}DM'B' \qquad (4.58)$$

The following two simultaneous matrix equations satisfy (4.54) identically

$$\dot{X}(t) = A_x X(t) + B_x Z(t) \qquad (4.59)$$

$$\dot{Z}(t) = A_z X(t) + B_z Z(t) \qquad (4.60)$$

Finally it is clear that the solutions to the linear matrix differential equations (4.59) and (4.60) with initial conditions $X(0) = P(0)$, $Z(0) = I$ when substituted into (4.52), satisfy the matrix Riccati equation given by (4.50). Since (4.59) and (4.60) are linear their solutions are unique. Hence there does indeed exist a solution $P(t)$ to (4.50) having the form (4.52) if $Z(t)$ is nonsingular.

Although this appears to be a suitable method for solving (4.50), it is in general a lengthy task to be carried out analytically, even when A_x, B_y, A_z, and B_z are constant matrices. If the solution is to be obtained by the aid of a computer, then it is just as well to use equation (4.50) itself. It should be noted that equation (4.50) does not depend on the observation $y(t)$. Consequently, $P(t)$ can be obtained before the start of the estimation process. If a digital computer is to be used for solving (4.50), then the matrix differential equation should be approximated by a matrix difference equation. This can be used as a justification for determining an approximate discrete model for the system (see Section 3.6) at the start, and then deriving a discrete optimal filter.

4.9 Continuous Estimator, Examples

In this section two examples are considered in detail where the procedure outlined in the preceding section is used to derive the optimum estimator.

▶ **example 1**

Let the signal $x(t)$ be a scalar random process with zero-mean and autocorrelation function

$$Ex(t_1)x(t_2) = \tfrac{3}{2}e^{-|t_2-t_1|}$$

We would like to derive the estimate $\hat{x}(t)$ minimizing $E[x(t) - \hat{x}(t)]^2$ by processing the observed data $y(\tau) = x(\tau) + v(\tau)$, $0 \leq \tau \leq t$ where $v(t)$ is white noise with correlation function

$$Ev(t_1)v(t_2) = \delta(t_2 - t_1)$$

The model for the process is given by

$$\begin{aligned}
\dot{x}(t) &= -x(t) + u(t) \\
y(t) &= x(t) + v(t) \\
Eu(t_1)u(t_2) &= 3\delta(t_2 - t_1) \\
Ev(t_1)v(t_2) &= \delta(t_2 - t_1) \\
Eu(t_1)v(t_2) &= 0
\end{aligned} \quad (4.61)$$

To obtain the steady-state random response, we must take the initial condition $x(0)$ as a random variable with variance p

$$Ex^2(0) = p = \tfrac{3}{2}$$

It is interesting, however, to leave p as a parameter and investigate the dependence of the estimator on p. Actually, if $x(0)$ was, for example, known to be zero, then p is equal to zero. From the model values for the terms in (4.55) through (4.58) we obtain

$$A_x = -1 \qquad B_x = 3 \qquad A_z = 1 \qquad B_z = 1$$

Hence, (4.59) and (4.60) become

$$\begin{aligned}
\dot{X}(t) &= -X(t) + 3Z(t) \qquad X(0) = P(0) = p \\
\dot{Z}(t) &= X(t) - Z(t) \qquad Z(0) = 1
\end{aligned} \quad (4.62)$$

These equations may be solved simultaneously by using the Laplace transformation

$$\begin{aligned}
sX(s) - X(0) &= -X(s) + 3Z(s) \\
sZ(s) - Z(0) &= X(s) + Z(s)
\end{aligned}$$

Solving for $X(s)$ yields

$$X(S) = \frac{ps - p + 3}{s^2 - 4}$$

or

$$X(t) = \tfrac{3}{4}(p - 1)e^{-2t} + \frac{p + 3}{4}e^{2t} \quad (4.63)$$

Solving for $Z(t)$ using (4.62) and $X(t)$ results in

$$Z(t) = \tfrac{1}{4}(1 - p)e^{-2t} + \tfrac{1}{4}(p + 3)e^{2t}$$

From (4.52)

$$P(t) = \frac{3(p - 1)e^{-2t} + (p + 3)e^{2t}}{(1 - p)e^{-2t} + (p + 3)e^{2t}} \qquad (4.64)$$

Since the terms $C'[DLD']^{-1}C$ and $C'[DLD']^{-1}$ in (4.51) are equal to unity here the optimum estimator is

$$\dot{\hat{x}} = -[1 + P(t)]\hat{x}(t) + P(t)y(t) \qquad (4.65)$$

where $P(t)$ is given by (4.64).

The asymptotic behavior of the optimal estimator can be studied as the observation interval $0 \leq \tau < t$ approaches infinity by determining the limit of $P(t)$

$$\lim_{t \to \infty} P(t) = 1$$

and (4.65) becomes

$$\dot{\hat{x}} = -2\hat{x}(t) + y(t)$$

which defines a filter with transfer function $\dfrac{1}{s + 2}$

$$\underrightarrow{y(t)} \boxed{\dfrac{1}{s + 2}} \underrightarrow{\hat{x}(t)}$$

It is interesting to note that the limit of $P(t)$ as $t \to \infty$ is independent of its initial value $P(0) = p$. This property holds in general when (4.27) is asymptotically stable or if the system (4.27) and (4.28) is completely observable (Section 3.8). These properties are readily verified for this ▶ example.

▶ **example 2**

In radar tracking of a nonmaneuvering (i.e., constant velocity) target a frequently encountered problem is to estimate the target position, say $x_1(t)$, and velocity* $\dot{x}_1(t) = x_2(t)$ from the received noisy position data $x_1(t) + v(t)$ where $v(t)$ is the additive white noise. A model for the system can be given by the following

$$\dot{x}(t) = \begin{bmatrix} 0 & 1 \\ 0 & 0 \end{bmatrix} x(t)$$

$$y(t) = [1 \quad 0]x(t) + v(t) \qquad (4.66)$$

$$Ev(t_1)v(t_2) = l\delta(t_2 - t_1)$$

$$Ex(0)x'(0) = P(0)$$

* For example, angular velocity in case of angle tracking.

Values for the terms in (4.55) through (4.58) are obtained from (4.66)

$$A_x = \begin{bmatrix} 0 & 1 \\ 0 & 0 \end{bmatrix}$$

$$B_x = 0$$

$$A_z = \begin{bmatrix} \dfrac{l}{1} & 0 \\ 0 & 0 \end{bmatrix} \qquad (4.67)$$

$$B_z = \begin{bmatrix} 0 & 0 \\ -1 & 0 \end{bmatrix}$$

Applying the Laplace transformation to (4.59) yields

$$sX(s) - X(0) = \begin{bmatrix} 0 & 1 \\ 0 & 0 \end{bmatrix} X(s)$$

or

$$X(s) = \left\{ sI - \begin{bmatrix} 0 & 1 \\ 0 & 0 \end{bmatrix} \right\}^{-1} P(0) \qquad (4.68)$$

Hence taking the inverse Laplace transform yields

$$X(t) = \begin{bmatrix} 1 & t \\ 0 & 1 \end{bmatrix} P(0)$$

The initial value $P(0)$ is to be taken as part of the available information. For example, if the initial position $x_1(0)$ and initial velocity $\dot{x}_1(0)$ are uniformly distributed over $(-\sqrt{2}, +\sqrt{2})$ and $(-4, +4)$ values respectively and if the initial position and velocity are uncorrelated then

$$Ex(0)x'(0) = P(0) = \begin{bmatrix} \frac{2}{3} & 0 \\ 0 & \frac{16}{3} \end{bmatrix}$$

Hence

$$X(t) = \begin{bmatrix} \frac{2}{3} & \frac{16}{3}t \\ 0 & \frac{16}{3} \end{bmatrix} \qquad (4.69)$$

Taking the Laplace transformation of (4.60) and using (4.67) and (4.68) and choosing a value for l such as $l = 1$ yields

$$sZ(s) - Z(0) = \begin{bmatrix} 0 & 0 \\ -1 & 0 \end{bmatrix} Z(s) + \begin{bmatrix} 1 & 0 \\ 0 & 0 \end{bmatrix} X(s)$$

$$Z(0) = I$$

Solving for $Z(s)$ results in

$$Z(s) = \begin{bmatrix} \dfrac{1}{s} & 0 \\[2ex] -\dfrac{1}{s^2} & \dfrac{1}{s} \end{bmatrix} + \begin{bmatrix} \dfrac{1}{s^2} & \dfrac{1}{s^3} \\[2ex] -\dfrac{1}{s^3} & -\dfrac{1}{s^4} \end{bmatrix} P(0)$$

Hence

$$Z(s) = \begin{bmatrix} \dfrac{1}{s} + \dfrac{\frac{2}{3}}{s^2} & \dfrac{\frac{16}{3}}{s^3} \\[3ex] -\dfrac{1}{s^2} - \dfrac{\frac{2}{3}}{s^3} & \dfrac{1}{s} - \dfrac{\frac{16}{3}}{s^4} \end{bmatrix}$$

or

$$Z(t) = \begin{bmatrix} 1 + \frac{2}{3}t & \frac{8}{3}t^2 \\[1ex] -t - \frac{1}{3}t^2 & 1 - \frac{1}{3}t^3 \end{bmatrix}$$

Consequently from (4.52)

$$P(t) = \begin{bmatrix} p_{11}(t) & p_{12}(t) \\ p_{12}(t) & p_{22}(t) \end{bmatrix}$$

where

$$p_{11}(t) = \frac{1}{\Delta} \left[\tfrac{2}{3} + \tfrac{16}{3}t^2 + \tfrac{32}{27}t^3 \right]$$

$$p_{12}(t) = \frac{1}{\Delta} \left[\tfrac{16}{3}t + \tfrac{16}{9}t^2 \right]$$

$$p_{22}(t) = \frac{1}{\Delta} \left[\tfrac{16}{3} + \tfrac{32}{9}t \right]$$

$$\Delta = 1 + \tfrac{2}{3}t + \tfrac{16}{9}t^3 - \tfrac{40}{27}t^4$$

and finally using (4.51) the optimal estimator is given by

$$\dot{\hat{x}}(t) = \begin{bmatrix} -p_{11}(t) & +1 \\ -p_{12}(t) & 0 \end{bmatrix} \hat{x}(t) + \begin{bmatrix} p_{11}(t) \\ p_{12}(t) \end{bmatrix} y(t)$$

It should be noted that in this example as t approaches infinity the covariance matrix $P(t)$ approaches zero and the optimal estimator becomes ▶ identical to the equation for $x(t)$ in (4.66).

4.10 Optimal Prediction, Continuous Case

The problem of estimating the value of $x(t + \delta)$; $\delta > 0$ given the observation $y(\tau)$, $0 < \tau \leq t$ is referred to as the problem of prediction. Linear prediction is then the case where we would like to determine $\alpha(\tau)$ such that

$$\hat{x}(t + \delta) = \int_0^t \alpha(\tau)y(\tau)\,d\tau \qquad (4.70)$$

minimizes the expected value of a positive definite function of estimation error such as the quadratic function

$$E[x(t + \delta) - \hat{x}(t + \delta)]Q[x(t + \delta) - \hat{x}(t + \delta)]$$

The choice of a quadratic function is again made to enable a tractable analysis.* All the terminology of Section 4.5 is assumed above.

Clearly following similar steps leading to equation (4.36) the principle of orthogonality now takes the form

$$E[x(t + \delta) - \hat{x}(t + \delta)]y'(\tau) = 0 \qquad 0 \le \tau \le t \qquad (4.71)$$

Solving (4.27) for $x(t + \delta)$ (equation (3.12)) yields

$$x(t + \delta) = X(t + \delta)\left\{x(0) + \int_0^{t+\delta} X^{-1}(\sigma)B(\sigma)u(\sigma)\,d\sigma\right\} \qquad (4.72)$$

which can be written as

$$x(t + \delta) = X(t + \delta)X^{-1}(t)X(t)$$
$$\times \left\{x(0) + \int_0^t X^{-1}(\sigma)B(\sigma)u(\sigma)\,d\sigma + \int_t^{t+\delta} X^{-1}(\sigma)B(\sigma)u(\sigma)\,d\sigma\right\}$$

since $X(t)$ is nonsingular. From (3.12) it then follows that

$$x(t + \delta) = X(t + \delta)X^{-1}(t)\left\{x(t) + \int_t^{t+\delta} X(t)X^{-1}(\sigma)B(\sigma)u(\sigma)\,d\sigma\right\} \quad (4.73)$$

Now the matrix $X(t)$ is the transition matrix, which is the nonsingular matrix satisfying

$$X(0) = I$$
$$\dot{X}(t) = AX(t)$$

Substituting for $x(t + \delta)$ from (4.73) into (4.71) and recalling that

$$Eu(\sigma)y'(\tau) = 0 \qquad \sigma > \tau$$

yields

$$E[X(t + \delta)X^{-1}(t)x(t) - \hat{x}(t + \delta)]y'(\tau) = 0 \qquad 0 \le \tau \le t \qquad (4.74)$$

The estimator for $\hat{x}(t + \delta)$ will be conjectured to have the form

$$\hat{x}(t + \delta) = X(t + \delta)X^{-1}(t)\hat{x}(t) \qquad (4.75)$$
$$\dot{\hat{x}}(t) = F_1\hat{x}(t) + F_2y(t) \qquad (4.76)$$

* If the process $x(t)$ is gaussian, minimization of this cost function will result in minimization of a class of functions, *See* Chapter 2.

Again, to verify this statement it is sufficient to determine the terms F_1 and F_2 such that (4.74) is satisfied. Substituting (4.75) into (4.74), and using the fact that $X(t + \delta)X^{-1}(t)$ is nonsingular yields

$$E[x(t) - \hat{x}(t)]y'(\tau) = 0 \qquad 0 \leq \tau \leq t \qquad (4.77)$$

Derivation of F_1 and F_2 in (4.76) satisfying (4.77) is the problem solved in Section 4.7 and the results are given by (4.40), (4.47), and (4.50).

In summary, the prediction problem, i.e., the determination of $\hat{x}(t + \delta)$, can be accomplished by the derivation of the estimate $\hat{x}(t)$ and then extrapolation of this estimate by use of the model of the random process. This is an intuitively logical result. The extrapolation is accomplished by multiplication of $\hat{x}(t)$ by $X(t + \delta)X^{-1}(t)$. In theory this multiplication is carried out instantaneously and, consequently the estimate $\hat{x}(t + \delta)$ is available at the same time as $\hat{x}(t)$.

When the matrix A in the model (4.27) is a constant matrix, then the term $X(t + \delta)X^{-1}(t)$ will take the form

$$X(t + \delta)X^{-1}(t) = e^{A(t+\delta)}e^{-At} = e^{A\delta} \qquad (4.78)$$

i.e., the extrapolation is only a function of the prediction interval δ. Now if $\delta < 0$, then we would have the so-called "interpolation problem." In this case, however, (4.74) is not valid, since $Eu(\sigma)y'(\tau)$ is not zero over the range of interest of σ and τ. Consequently, (4.75) and (4.76) cannot be used to yield $\hat{x}(t + \delta)$ when δ is negative. It is, however, possible to solve the interpolation problem by methods similar to those presented herein (see Section 4.13).

4.11   Optimal Prediction, Discrete Case

Discrete optimum prediction is the problem of estimating $x(k + \delta)$, $\delta > 1$ given $y(i)$, $0 \leq i \leq k$ where the quantities k, δ, and i are integers. Using a procedure identical to that used in Section 4.10, one can show that the optimal estimate $\hat{x}(k + \delta)$ can be obtained from $\hat{x}(k + 1)$ by extrapolating it through the model of the discrete process, i.e., the solution of the difference equation

$$x(j + 1) = A(j)x(j) \qquad (4.79)$$

by writing

$$\hat{x}(k + \delta) = A(k + \delta - 1) \cdots A(k + 2)A(k + 1)\hat{x}(k + 1) \qquad (4.80)$$

In case $A(j)$ is a constant matrix (independent of j) then

$$\hat{x}(k + \delta) = A^{(\delta-1)}\hat{x}(k + 1) \qquad (4.81)$$

Finally the derivation of the optimum estimate $\hat{x}(k + 1)$ given $y(0), \ldots, y(k)$ is the problem discussed in Section 4.2.

4.12 The Problem of Missing Data

Due to component failure or other similar reasons it is possible in practice that either no observation is made available over an interval say $t_1 < t \leq t_2$ (or $k_1 < k \leq k_2$) or the observations received are highly unreliable and have to be discarded. The optimal procedure for estimation of $x(t)$, $t_1 < t \leq t_2$ is clearly evident. We have to use the prediction scheme discussed in Section 4.10 (or 4.11 in the discrete case) in order to continually estimate $x(t)$ for $t > t_1$ using the last available optimal estimate $\hat{x}(t_1)$ until the observation resumes again (e.g., at $t = t_2$). From the practical point of view this amounts to setting $y(t) = 0$ and $P(t)C'[DLD']^{-1}C = 0$ in the estimator equation (4.51) at $t = t_1$. The normal operation of (4.51) resumes again at t_2. However, at $t = t_2$ we have an optimal estimate $\hat{x}(t_2)$ and the value for $P(t_2)$ should be chosen accordingly.* The discrete case can be handled in an exactly similar manner.

4.13 Optimal Interpolation, Discrete Case

Given the observations $y(i)$, $i = 0, \ldots, k$, we may be interested in determining the best estimate of the state $x(k - j)$ for some $j > 0$. This is referred to as the interpolation problem. Let us denote such an estimate by

$$\hat{x}^{(j)}(k - j)$$

This estimate is also referred to as the j-step interpolated value of $x(k - j)$. The problem of optimal interpolation is the determination of constants (in general matrices) $\alpha(i)$, $i = 0, \ldots, k$ such that the linear estimate given by

$$\hat{x}^{(j)}(k - j) = \sum_0^k \alpha(i)y(i) \tag{4.82}$$

minimizes the quadratic cost function

$$E[x(k - j) - \hat{x}^{(j)}(k - j)]'Q[x(k - j) - \hat{x}^{(j)}(k - j)]$$

for any positive definite matrix Q. Substituting (4.82) and differentiating with respect to $\alpha(i)$, we arrive at the orthogonality principle given by

$$E[x(k - j) - \hat{x}^{(j)}(k - j)]y'(i) = 0 \qquad i = 0, \ldots, k \tag{4.83}$$

We will use the notation $\hat{x}(i + 1)$ to represent a one-step prediction of $x(i + 1)$ as in (4.7). It is conjectured that, given the fixed observation sequence $y(i)$, $i = 0, \ldots, k$ the jth optimal interpolation can be given as the recursive formula

$$\hat{x}^{(j)}(k - j) = F_1\hat{x}^{(j-1)}(k - j + 1) + F_2\hat{x}(k - j + 1) \tag{4.84}$$

where $\hat{x}(k - j + 1)$ represents the one-step prediction of $x(k - j + 1)$ using the observations $y(i)$, $i = 0, \ldots, k - j$ and where $\hat{x}^{(j-1)}(k - j + 1)$ is

* *See* the discussion following (4.51).

the $(j-1)$st optimal interpolation given the observation $y(i)$; $i = 0, \ldots, k$. Equation (4.84) is a difference equation in $\hat{x}^{(j)}(k-j)$, with k constant and j the variable index $(j = 0, 1, \ldots)$. This equation has a forcing function which is linear in the one-step predicted value $\hat{x}(k-j+1)$. As an example, (4.84) implies that if we have received 10 observations $y(0), \ldots, y(9)$ the best estimate of $x(5)$ given $y(0), \ldots, y(9)$ is obtained from the best estimate of $x(6)$ given $y(0), \ldots, y(5)$ (this is given by the estimator derived in Section 4.4) and the best estimate of $x(6)$ given $y(0), \ldots, y(9)$.

As before, in order to verify that (4.84) is a meaningful relationship it is sufficient to show that values for F_1 and F_2 can be obtained such that the orthogonality principle (4.83) is satisfied.

By substituting (4.84) into (4.83) it follows that

$$E[x(k-j) - F_1\hat{x}^{(j-1)}(k-j+1) - F_2\hat{x}(k-j+1)]y'(i) = 0,$$
$$i = 0, \ldots, k \quad (4.85)$$

Since $\hat{x}^{(j-1)}(k-j+1)$ is assumed to be the optimal estimate of $x(k-j+1)$ given $y(i), i = 0, \ldots, k$, it must satisfy

$$E[x(k-j+1) - \hat{x}^{(j-1)}(k-j+1)]y'(i) = 0, \quad i = 0, \ldots, k \quad (4.86)$$

where (4.86) is obtained from (4.83) by substituting $j-1$ for j. Multiplying (4.86) by F_1 and substituting into (4.85) yields

$$E[x(k-j) - F_1x(k-j+1) - F_2\hat{x}(k-j+1)]y'(i) = 0,$$
$$i = 0, \ldots, k \quad (4.87)$$

From (4.1) we have

$$x(k-j+1) = A(k-j)x(k-j) + B(k-j)u(k-j) \quad (4.88)$$

Equation (4.88) can be solved for $x(k-j)$ if A is nonsingular.* It follows that

$$x(k-j) = A^{-1}(k-j)x(k-j+1) - A^{-1}(k-j)B(k-j)u(k-j) \quad (4.89)$$

Substituting (4.89) into (4.87), it follows that

$$E\{[A^{-1}(k-j) - F_1]x(k-j+1) - F_2\hat{x}(k-j+1)$$
$$- A^{-1}(k-j)B(k-j)u(k-j)\}y'(i) = 0, \quad i = 0, \ldots, k \quad (4.90)$$

Let us add and subtract the term $[A^{-1}(k-j) - F_1]\hat{x}(k-j+1)$ within the bracket in (4.90). Using the notation $\tilde{x} = x - \hat{x}$ it follows that

$$E\{[A^{-1}(k-j) - F_1]\tilde{x}(k-j+1)$$
$$+ [A^{-1}(k-j) - F_1 - F_2]\hat{x}(k-j+1)$$
$$- A^{-1}(k-j)B(k-j)u(k-j)\}y'(i) = 0, \quad i = 0, \ldots, k \quad (4.91)$$

* A is nonsingular if (4.1) is the discrete representation of a differential equation. *See* Section 3.6.

The set of equations given by (4.91) can be decomposed into two groups, the first, corresponding to $i = 0, \ldots, k - j$ and the second corresponding to $i = k - j + 1, \ldots, k$.

For the first group we have

$$E\{[A^{-1}(k - j) - F_1]\tilde{x}(k - j + 1)$$
$$+ [A^{-1}(k - j) - F_1 - F_2]\hat{x}(k - j + 1)$$
$$- A^{-1}(k - j)B(k - j)u(k - j)\}y'(i) = 0 \qquad i = 0, \ldots, k - j \quad (4.92)$$

Since $\hat{x}(k - j + 1)$ is by definition the optimal estimate given the observations $y(0), \ldots, y(k - j)$, we know from the orthogonality principle that

$$E\tilde{x}(k - j + 1)y'(i) = 0; \qquad i = 0, \ldots, k - j \qquad (4.93)$$

Using (4.93) and the fact that $u(k - j)$ is independent of $y(i)$ for $i \leq k - j$, (4.92) reduces to

$$E[A^{-1}(k - j) - F_1 - F_2]\hat{x}(k - j + 1)y'(i) = 0,$$
$$i = 0, \ldots, k - j \qquad (4.94)$$

Now (4.94) can be satisfied if

$$A^{-1}(k - j) - F_1 - F_2 = 0 \qquad (4.95)$$

Equation (4.95) is one relationship between F_1 and F_2. We now search for another relationship satisfying the second group of equations obtained by decomposition of (4.91), namely

$$E\{[A^{-1}(k - j) - F_1]\tilde{x}(k - j + 1) - A^{-1}(k - j)B(k - j)u(k - j)\}y'(i) = 0,$$
$$i = k - j + 1, \ldots, k \qquad (4.96)$$

where we have used the relationship given by (4.95).

Let us first consider $i = k - j + 1$. From (4.96) and (4.2) it follows that

$$E\{[A^{-1}(k - j) - F_1]\tilde{x}(k - j + 1) - A^{-1}(k - j)B(k - j)u(k - j)\}$$
$$\times [Cx(k - j + 1) + Dv(k - j + 1)]' = 0 \quad (4.97)$$

The term $Dv(k - j + 1)$ can be dropped since it is clearly independent of each term in the left hand bracket. Furthermore, since $\hat{x}(k - j + 1)$ is linear in $y(i)$, $i = 0, \ldots, k - j$ and using (4.93) then equation (4.97) can be written

$$E\{[A^{-1}(k - j) - F_1]\tilde{x}(k - j + 1)\}\hat{x}'(k - j + 1)C'$$
$$- E[A^{-1}(k - j)B(k - j)u(k - j)x'(k - j + 1)C'] = 0$$

Recalling the definition

$$P(k) = Ex(k)x'(k)$$

it follows that

$$[A^{-1}(k-j) - F_1]P(k-j+1)C'$$
$$- E[A^{-1}(k-j)B(k-j)u(k-j)x'(k-j+1)C'] = 0 \quad (4.98)$$

From (4.88) and since $u(k-j)$ is independent of $x(k-j)$ it follows that

$$\{[A^{-1}(k-j) - F_1]P(k-j+1)$$
$$- A^{-1}(k-j)B(k-j)KB'(k-j)\}C' = 0 \quad (4.99)$$

Finally, (4.99) is satisfied if we take

$$F_1 = Q(k-j) \quad (4.100)$$

where we have defined

$$Q(i) = A^{-1}(i)[I - B(i)KB'(i)P^{-1}(i+1)] \quad (4.101)$$

In the following it will be shown that the condition (4.100) will cause equation (4.96) to be satisfied for $i = k - j + 2, \ldots, k$ also. From (4.100) and (4.101) it follows that

$$A^{-1}(k-j) - F_1 = A^{-1}(k-j)B(k-j)KB'(k-j)P^{-1}(k-j+1)$$

Substituting into (4.96) and cancelling a factor $A^{-1}(k-j)$ yields the requirement that

$$E\{B(k-j)KB'(k-j)P^{-1}(k-j+1)x(k-j+1)$$
$$- B(k-j)u(k-j)\}y'(i) = 0, \quad i = k - j + 1, \ldots, k \quad (4.102)$$

Let us choose any value of i such as l so that $k - j + 1 \leq l \leq k$. We then have

$$y(l) = Cx(l) + Dv(l) \quad (4.103)$$

Furthermore, solving the difference equation (4.1) for $x(l)$ with the initial condition $x(k-j+1)$ yields (*see* Chapter 1)

$$x(l) = Mx(k-j+1) + \text{a function of } u(i), \ i \geq k-j+1$$

Therefore (4.102) for $i = l$ can be written as

$$EB(k-j)KB'(k-j)P^{-1}(k-j+1)x(k-j+1)x'(k-j+1)M'C'$$
$$- EB(k-j)u(k-j)x'(k-j+1)M'C' = 0 \quad (4.104)$$

Finally, recalling that

$$E\tilde{x}x' = E\tilde{x}[x' - \hat{x}'] = E\tilde{x}\tilde{x}' = P$$

and

$$Eu(k-j)x'(k-j+1) = KB'(k-j)$$

it is seen that (4.104) is satisfied as an identity. This completes the proof of the assertion concerning existence of F_1 and F_2 satisfying (4.85). In summary from (4.85), (4.95), and (4.100) we have

$$\hat{x}^{(j)}(k - j) = Q(k - j)\hat{x}^{(j-1)}(k - j + 1)$$
$$+ [A^{-1}(k - j) - Q(k - j)]\hat{x}(k - j + 1) \quad (4.105)$$

where

$$Q(i) = A^{-1}(i)[I - B(i)KB'(i)P^{-1}(i + 1)] \quad (4.106)$$

A discussion of equations (4.105) and (4.106) is warranted here. When we have received the observations $y(0), \ldots, y(k)$, the results of Section 4.4 yield the value of $\hat{x}(k + 1)$, i.e., the corresponding one-step prediction. We should also store the values of $\hat{x}(k)$, $\hat{x}(k - 1), \ldots, \hat{x}(k - j + 1)$ along with the values of $P(k)$, $P(k - 1), \ldots, P(k - j + 1)$. These values of the covariance matrix have already been obtained by the time the kth observation arrives in the process of determining the estimate $\hat{x}(k + 1)$. Using these and starting with $\hat{x}^{(0)}(k)$ equation (4.105) is recursively solved to obtain $\hat{x}^{(1)}(k - 1)$, $\hat{x}^{(2)}(k - 2), \ldots$, down to $\hat{x}^{(j)}(k - j)$, which is the desired result. The interpretation of $\hat{x}^{(0)}(k)$ and an example of an interpolation problem are given in the following two sections.

4.14   Discrete Filtering

The filtering problem is the determination of the estimate of $x(k)$ given the observation $y(0), \ldots, y(k)$. According to the notation (4.85) of the preceding section, the filtered value of $x(k)$ is denoted by

$$\hat{x}^{(0)}(k) \quad (4.107)$$

This can be considered as a special case of the interpolation problem* if we interpret $\hat{x}^{(-1)}(k + 1)$ as $\hat{x}(k + 1)$, which is the one-step predictor of $x(k + 1)$. Returning to (4.105), and setting $j = 0$, and using

$$\hat{x}^{(-1)}(k + 1) = \hat{x}(k + 1)$$

it follows that

$$\hat{x}^{(0)}(k) = A^{-1}(k)\hat{x}(k + 1) \quad (4.108)$$

In other words, the filtered value of the kth sample $(x(k))$ is equal to $A^{-1}(k)$ multiplied by its one-step predicted value if the observation is $y(0), \ldots, y(k)$. Equation (4.108) can be verified directly. By definition $\hat{x}^{(0)}(k)$ should satisfy the orthogonality condition

$$E[x(k) - \hat{x}^{(0)}(k)]y'(i) = 0 \qquad i = 0, \ldots, k \quad (4.109)$$

* This approach requires, however, that we show validity of (4.105) for $j = 0$.

Substituting for $\hat{x}^{(0)}(k)$ from (4.108) and multiplying by $A(k)$ yields

$$E[A(k)x(k) - \hat{x}(k+1)]y'(i) = 0 \qquad i = 0, \ldots, k \qquad (4.110)$$

Since

$$Eu(k)y'(i) = 0 \qquad i = 0, \ldots, k$$

equation (4.110) is equivalent to

$$E[A(k)x(k) + B(k)u(k) - \hat{x}(k+1)]y'(i) = 0 \qquad i = 0, \ldots, k \qquad (4.111)$$

Finally, from (4.1) it follows that

$$E[x(k+1) - \hat{x}(k+1)]y'(i) = 0 \qquad i = 0, \ldots, k \qquad (4.112)$$

Since (4.112) is just the orthogonality condition that $\hat{x}(k+1)$ should satisfy, it follows that (4.109) is satisfied when the filter has the form (4.108).

4.15 Examples

▶ **example 1**

First let us consider Example 1 of Section 4.4. We have

$$A(k-j) = 1$$

$$B(i) = 0$$

Hence from (4.106)

$$Q(i) = 1$$

and from (4.105)

$$\hat{x}^{(j)}(k-j) = \hat{x}^{(j-1)}(k-j+1)$$

In other words, given the observation $y(0), \ldots, y(k)$, the one-step predicted value* $\hat{x}(k+1)$ is the same as the filtered value $\hat{x}(k)$ and all interpolated (smoothed) values $\hat{x}^{(j)}(k-j)$, $1 \leq j \leq k$. This is intuitively an acceptable result since we are trying to estimate a quantity which is known *a priori* to be a constant [the mean of the random sequence $y(k)$].

In fact, from (4.106) we observe that if $B(i) = 0$, that is, the parameter we are to estimate is a deterministic quantity and the randomness is only due to the observation noise, then

$$Q(i) = A^{-1}(i)$$

and from (4.105)

$$\hat{x}^{(j)}(k-j) = A^{-1}(k-j)\hat{x}^{(j-1)}(k-j+1) \qquad (4.113)$$

Equation (4.113) implies that, in this case, any degree of interpolation is obtained by "diffusion" of $\hat{x}(k+1)$ backward through the system $x(k+1) = Ax(k)$.

---

* In fact for any predicted value of $x(k+s)$, $s \geq 1$.

▶ **example 2**

Finally, let us consider the following scalar problem.

$$x(k + 1) = x(k) + u(k)$$
$$y(k) = x(k) + v(k)$$
$$Eu(k_1)u(k_2) = Ev(k_1)v(k_2) = \Delta(k_2 - k_1) \qquad (4.114)$$
$$Eu(k_1)v(k_2) = 0$$

Hence the parameters in (4.24–4.26) are given by

$$A = B = C = D = K = L = 1$$

and the equations for the one-step prediction (4.24–4.26) become

$$\hat{x}(k + 1) = [1 - F(k)]\hat{x}(k) + F(k)y(k)$$
$$F(k) = \frac{P(k)}{1 + P(k)} \qquad (4.115)$$
$$P(k + 1) = [1 - F(k)]^2 P(k) + 1 + F^2(k)$$

Given the observations say, $y(0), \ldots, y(10)$, and $P(0)$ these equations yield

$$\hat{x}(1), \hat{x}(2), \ldots, \hat{x}(11) \qquad (4.116)$$

We also have available

$$P(1), P(2), \ldots, P(k), \ldots \qquad (4.117)$$

Suppose we would like to determine the best estimate of $x(9)$. Then we need the values of

$$\hat{x}(11) = \alpha, \qquad \hat{x}(10) = \beta, \qquad P(10) = \gamma \qquad (4.118)$$

From (4.114) equations (4.105) and (4.106) become

$$\hat{x}^{(j)}(10 - j) = P^{-1}(10 + j - 1)\hat{x}(10 - j + 1)$$
$$+ [1 - P^{-1}(10 + j - 1)]\hat{x}^{(j-1)}(10 - j + 1)$$
$$Q(i) = 1 - P^{-1}(i + 1) \qquad (4.119)$$

Starting with $\hat{x}(11) = \alpha$, we have from (4.108) that

$$\hat{x}^{(0)}(10) = A^{-1}\hat{x}(11) = \hat{x}(11) = \alpha$$

Hence, (4.119) yields

▶ $$\hat{x}^{(1)}(9) = P^{-1}(10)\hat{x}(10) + [1 - P^{-1}(10)]\hat{x}^{(0)}(10) = \frac{\beta}{\gamma} + \left(1 - \frac{1}{\gamma}\right)\alpha$$

4.16 A Special Case

The solution to the optimal interpolation problem was obtained in Section 4.12 in the form of a backward recursive equation relating the interpolated value of the sample $k - j$ to that of sample $k - j + 1$, where k is fixed during this recursive process and corresponds to the last received observation. A more desirable relationship, as in the case of the one-step prediction of Section 4.4, is a recursive relationship between the best estimate based on the kth observation and the best estimate based on the $(k - 1)$th observation plus the kth observation. In other words, we would like to use the last observation to construct an additive "updating" term to add to the previous best estimate and result in the new best estimate, and, consequently have no need to store any of the past estimates. Unfortunately such a form does not exist for the general problem considered in Section 4.12.

However, in practice there are cases where we may be interested in the interpolated value corresponding to a fixed sample independent of the length of observation. For example, we may be interested in the best (*a posteriori*) estimate of the initial condition of (4.1) for any sequence of observations $y(0), \ldots , y(k)$, any k. This situation arises, for instance, after the launching of a rocket, if we desire to determine the accuracy of the launch parameters such as initial acceleration and velocity, by processing the information received by tracking the vehicle trajectory. It will be seen in the following that for this particular case the estimator assumes a more desirable form.

Let $x(p)$, where p is fixed [not changing with the length of received observation sequence $y(0), \ldots , y(k)$] be the parameter to be estimated. Let*

$$k - p = j, \quad k > p \tag{4.120}$$

Therefore the value $x(p)$ is the same as $x(k - j)$ and the optimal estimate of $x(p)$ given $y(0), \ldots , y(k)$ is

$$\hat{x}^{(j)}(k - j) = \hat{x}^{(j)}(p) \tag{4.121}$$

where $\hat{x}^{(j)}(p)$ denotes the estimate of $x(p)$ obtained through the jth step of interpolation (smoothing) implying that there should then be $j + p$ observations available. Equation (4.105) can now be written as

$$\hat{x}^{(j)}(p) = [A^{-1}(p) - Q(p)]\hat{x}(p + 1) + Q(p)\hat{x}^{(j-1)}(p + 1) \tag{4.122}$$

Substituting $j - 1$ for j in (4.122) we obtain the following value for the $(j - 1)$th smoothed value of $x(p)$ [i.e., the interpolation of $x(p)$ corresponding to the observations $y(0), \ldots , y(k - 1)$].

$$\hat{x}^{(j-1)}(p) = [A^{-1}(p) - Q(p)]\hat{x}(p + 1) + Q(p)\hat{x}^{(j-2)}(p + 1) \tag{4.123}$$

Subtracting (4.123) from (4.122) yields

$$\hat{x}^{(j)}(p) - \hat{x}^{(j-1)}(p) = Q(p)[\hat{x}^{(j-1)}(p + 1) - \hat{x}^{(j-2)}(p + 1)] \tag{4.124}$$

---

* For $k = p$ we have the filtering problem discussed in Section 4.13 and for $k < p$ we are dealing with the discrete prediction problem discussed in Section 4.11. The corresponding estimates for these cases already have the desirable recursive form.

Similarly if we start with $\hat{x}^{(j-1)}(p+1)$ in place of $\hat{x}^{(j)}(p)$ it follows that

$$\hat{x}^{(j-1)}(p+1) - \hat{x}^{(j-2)}(p+1)$$
$$= Q(p+1)[\hat{x}^{(j-2)}(p+2) - \hat{x}^{(j-3)}(p+2)] \quad (4.125)$$

Substituting (4.125) into (4.124) and repeating this process it follows that

$$\hat{x}^{(j)}(p) - \hat{x}^{(j-1)}(p) = \left[\prod_{p}^{k-1} Q(i) \right][\hat{x}^{(0)}(k) - \hat{x}(k)] \quad (4.126)$$

From (4.108), i.e., the relationship for the filtered value of $x(k)$, we have

$$\hat{x}^{(0)}(k) - \hat{x}(k) = A^{-1}(k)\hat{x}(k+1) - \hat{x}(k) \quad (4.127)$$

Substituting for $\hat{x}(k+1)$, the one-step predicted value, from (4.24) yields

$$\hat{x}^{(0)}(k) - \hat{x}(k) = A^{-1}(k)F(k)[y(k) - C\hat{x}(k)] \quad (4.128)$$

Finally substituting into (4.126) yields

$$\hat{x}^{(j)}(p) = \hat{x}^{(j-1)}(p) + \left[\prod_{p}^{k-1} Q(i) \right]A^{-1}(k)F(k)[y(k) - C\hat{x}(k)] \quad (4.129)$$

where $F(k)$ is given by (4.25) and $Q(i)$ is given by (4.106). The use of equation (4.129) is illustrated as follows. Given the observations up to $k-1$ and having established the corresponding $(j-1)$th smoothed value of $x(p)$, we construct the additive correction to the $(j-1)$th smoothed value by using the observation $y(k)$. We also need the value $\hat{x}(k)$ which suggests that we need to implement a usual one-step predictor. The coefficient of $[y(k) - C\hat{x}(k)]$ can either be computed before any observation is received or during the estimation process. Again, this decision is based on whether we prefer to use additional memory in place of more "on-line" computations.

▶ **example**
Let us consider a stochastic system described by (4.114), where we desire to determine the best estimate of the initial condition $x(0)$ by processing the observation $y(0), \dots, y(k)$. Equation (4.129) yields

$$\hat{x}^{(k)}(0) = \hat{x}^{(k-1)}(0) + \left[\prod_{0}^{k-1} Q(i) \right]A^{-1}(k)F(k)[y(k) - C\hat{x}(k)] \quad (4.130)$$

where $\hat{x}^{(k)}(0)$ is the kth smoothed (interpolated) value of the initial condition $x(0)$, i.e., the best estimate of $x(0)$ given $y(0), \dots, y(k)$. Using (4.114) and (4.119) we have

$$Q(i) = 1 - P^{-1}(i+1)$$
$$A(k) = 1$$
$$C = 1$$

Therefore, (4.130) becomes

$$x^{(k)}(0) = \hat{x}^{(k-1)}(0) + \left\{ \prod_0^{k-1} [1 - P^{-1}(i + 1)] \right\} F(k)[y(k) - \hat{x}(k)]$$

$$(4.131)$$

This is the recursive equation of the interpolator. The estimate $\hat{x}(k)$ (one-step prediction) and the parameter $F(k)$ are obtained by using (4.24–4.26). ▶ $F(k)$ is given explicitly by (4.115).

4.17   Optimal Interpolation, Continuous Case

The continuous version of the interpolation problem discussed in Section 4.13 is introduced as follows. Given the observation $y(\tau)$, $0 \leq \tau \leq t$, we are interested in determining the best mean square estimate of the state $x(t - \sigma)$ for some $\sigma > 0$. Let us denote this estimate by

$$\hat{x}^{(\sigma)}(t - \sigma)$$

Note that this estimate is available at time t. Two specific problems are solved in this section. First, when σ is a given constant and t is a running variable. This implies that, at any time t, we wish to estimate the value of the state which existed σ seconds in the past. The second problem is when we wish to estimate the state at a fixed time, such as $x(a)$ where the observed data $y(\tau)$ are given over an interval $0 \leq \tau \leq t$, $(t > a)$. In this case, following the above notation, the estimate of $x(a)$ is denoted by $\hat{x}^{(\sigma)}(a)$ where now a is a constant and σ is a variable $(\sigma = t - a)$. It is observed that the second problem is a special case of the first problem. The orthogonality principle [equation (4.36)] now takes the form

$$E[x(t - \sigma) - \hat{x}^{(\sigma)}(t - \sigma)]y'(\tau) = 0 \qquad 0 \leq \tau \leq t \qquad (4.132)$$

In order to derive (4.132) all that is necessary is to replace $x(t)$ and $\hat{x}(t)$ in (4.32) and (4.33) by $x(t - \sigma)$ and $\hat{x}^{(\sigma)}(t - \sigma)$, respectively and follow the steps leading to (4.36).

Using a treatment similar to that used for optimal interpolation in the discrete case, here we conjecture that a formulation for the interpolation problem in the continuous case can be obtained in the form

$$\frac{d\hat{x}^{(\sigma)}(t - \sigma)}{d\sigma} = F_1 \hat{x}^{(\sigma)}(t - \sigma) + F_2 \hat{x}(t - \sigma) \qquad (4.133)$$

i.e., for a given t and a *variable* σ, the optimal interpolated value $\hat{x}^{(\sigma)}(t - \sigma)$ satisfies a linear differential equation with a forcing function composed of $\hat{x}(t - \sigma)$, the optimal filtered value.* The variable σ is positive and varies

---

* $\hat{x}(t - \sigma)$ is the best estimate of $x(t - \sigma)$ given the observation $y(\tau)$, $0 \leq \tau \leq t - \sigma$. Its value is obtained from the results of Section 4.7.

between zero and t. It may be seen that (4.133) is the continuous analog of (4.85). Again, to verify that (4.133) is a valid formulation for the continuous optimal interpolation problem it is sufficient to show that matrices F_1 and F_2 exist such that (4.133) satisfies the orthogonality principle (4.132). By definition, the initial condition for (4.133) is given by

$$\hat{x}^{(0)}(t) = \hat{x}(t)$$

Consequently, from (4.36) it is clear that the orthogonality condition (4.132) is satisfied at $\sigma = 0$.

The equation (4.132) can be decomposed into three parts, namely,

$$E[x(t - \sigma) - \hat{x}^{(\sigma)}(t - \sigma)]y'(\tau) = 0 \qquad 0 \leq \tau < t - \sigma \qquad (4.134)$$

$$E[x(t - \sigma) - \hat{x}^{(\sigma)}(t - \sigma)]y'(\tau) = 0 \qquad t - \sigma < \tau \leq t \qquad (4.135)$$

$$E[x(t - \sigma) - \hat{x}^{(\sigma)}(t - \sigma)]y'(t - \sigma) = 0 \qquad (4.136)$$

Let us differentiate (4.134) with respect to σ. It follows that

$$E\left[\frac{d}{d\sigma} x(t - \sigma) - \frac{d}{d\sigma} \hat{x}^{(\sigma)}(t - \sigma)\right]y'(\tau) = 0 \qquad 0 \leq \tau < t - \sigma \quad (4.137)$$

Since $x(t)$ satisfies the differential equation (4.27), we have

$$\frac{dx(\gamma)}{d\gamma} = Ax(\gamma) + Bu(\gamma) \qquad 0 \leq \gamma$$

where γ is a dummy variable which is substituted for t. We can use the change of variable $\gamma = t - \sigma$, where t is a constant and σ is the variable. It follows that

$$\frac{d}{d\sigma} x(t - \sigma) = -Ax(t - \sigma) - Bu(t - \sigma) \qquad \sigma \leq t \qquad (4.138)$$

Substituting (4.133) and (4.138) into (4.137) yields

$$E[-Ax(t - \sigma) - Bu(t - \sigma) - F_1\hat{x}^{(\sigma)}(t - \sigma) - F_2\hat{x}(t - \sigma)]$$
$$\times y'(\tau) = 0 \qquad 0 \leq \tau < t - \sigma \quad (4.139)$$

Let us add and subtract the term $(F_1 + F_2)x(t - \sigma)$ within the bracket in (4.139). Utilizing the orthogonality condition satisfied by $\hat{x}(t - \sigma)$ [equation (4.36) when we substitute $t - \sigma$ for t] and (4.134) and the fact that

$$Eu(t - \sigma)y'(\tau) = Eu(t - \sigma)[Cx(\tau) + Dv(\tau)]' = 0$$
$$0 \leq \tau < t - \sigma \quad (4.140)$$

it follows that

$$E[-A - F_1 - F_2]x(t - \sigma)y'(\tau) = 0 \qquad 0 \leq \tau < t - \sigma \quad (4.141)$$

Consequently (4.141), and hence (4.137), is satisfied if we choose F_1 and F_2 so that

$$F_1 + F_2 = -A \qquad (4.142)$$

Differentiating (4.135) with respect to σ yields

$$E\left[\frac{d}{d\sigma}\, x(t - \sigma) - \frac{d}{d\sigma}\, \hat{x}^{(\sigma)}(t - \sigma)\right]y'(\tau) = 0 \qquad t - \sigma < \tau \le t \quad (4.143)$$

Substituting for the terms in the bracket from (4.138) and (4.133) and utilizing (4.142) and (4.135) yields

$$E[-Bu(t - \sigma) + F_2\tilde{x}(t - \sigma)]y'(\tau) = 0 \qquad t - \sigma < \tau \le t \quad (4.144)$$

where

$$\tilde{x}(t - \sigma) \triangleq x(t - \sigma) - \hat{x}(t - \sigma)$$

The observation $y(\tau)$, using (4.27) and (4.28) can be written as follows

$$y(\tau) = C\{X(\tau)x(0) + \int_0^\tau X(\tau)X^{-1}(s)Bu(s)\,ds\} + Dv(\tau); \; t - \sigma < \tau \le t \quad (4.145)$$

Therefore assuming $x(0)$ and $u(t - \sigma)$ are uncorrelated we have

$$Eu(t - \sigma)y'(\tau) = \int_0^\tau Eu(t - \sigma)u'(s)B'X^{-1'}(s)X'(\tau)C'\,ds$$
$$= KB'(t - \sigma)X^{-1'}(t - \sigma)X'(\tau)C' \quad (4.146)$$

Furthermore, since

$$E\tilde{x}(t - \sigma)u'(s) = 0 \qquad s > t - \sigma$$
$$E\tilde{x}(t - \sigma)v'(\tau) = 0 \qquad \tau > t - \sigma$$

and utilizing (4.145), we have

$$E\tilde{x}(t - \sigma)y'(\tau) = E\tilde{x}(t - \sigma)\left[X(\tau)x(0) + \int_0^{t-\sigma} X(\tau)X^{-1}(s)Bu(s)\,ds\right]'C'$$
$$\tau > t - \sigma \quad (4.147)$$

The quantity in the bracket in the right hand side of (4.147) can be written

$$X(\tau)x(0) + \int_0^{t-\sigma} X(\tau)X^{-1}(s)Bu(s)\,ds = X(\tau)X^{-1}(t - \sigma)x(t - \sigma)$$

Substituting the above into (4.147) and utilizing*

$$E\tilde{x}(t - \sigma)\hat{x}(t - \sigma) = 0$$

it follows that

$$E\tilde{x}(t - \sigma)y'(\tau) = E\tilde{x}(t - \sigma)\tilde{x}'(t - \sigma)X^{-1'}(t - \sigma)X'(\tau)C' \quad (4.148)$$

* Recall that $\hat{x}(t - \sigma)$ is a linear function of $y(\tau)$, $\tau \le t - \sigma$.

Finally substituting (4.146) and (4.148) into (4.144) yields

$$-BKB'X^{-1'}(t - \sigma)X'(\tau)C' + F_2P(t - \sigma)X^{-1'}(t - \sigma)X'(\tau)C' = 0 \quad (4.149)$$

Equation (4.149), and hence (4.143), is then satisfied if we choose

$$F_2 \triangleq F_2(t - \sigma) = BKB'P^{-1}(t - \sigma) \quad (4.150)$$

Since the expression in the left hand side of (4.132) is continuous and differentiable it is easy to show that the remaining condition [equation (4.136)] is also satisfied if F_1 and F_2 are chosen according to (4.142) and (4.150).

The final results thus far obtained for the optimal interpolator are repeated below

$$\frac{d\hat{x}^{(\sigma)}(t - \sigma)}{d\sigma} = [-A - BKB'P^{-1}(t - \sigma)]\hat{x}^{(\sigma)}(t - \sigma)$$

$$+ BKB'P^{-1}(t - \sigma)\hat{x}(t - \sigma) \quad (4.151)$$

where

$$P(t - \sigma) = E\tilde{x}(t - \sigma)\tilde{x}'(t - \sigma)$$

$$Eu(t_1)u'(t_2) = K\,\delta(t_2 - t_1)$$

Equation (4.151) is the desired result. The second problem introduced at the beginning of the section may now be solved. Clearly the interpolated value $\hat{x}^{(\sigma)}(t - \sigma)$ can be obtained as the solution of (4.151) by integrating (4.151) with respect to σ, i.e., for a given t and σ we have

$$\hat{x}^{(\sigma)}(t - \sigma) = X(t - \sigma)X^{-1}(t)\hat{x}(t) - \int_t^{t-\sigma} X(t - \sigma)X^{-1}(\delta)F_2(\delta)\hat{x}(\delta)\,d\delta \quad (4.152)$$

where $X(t)$ now represents the transition matrix associated with (4.151) and is defined by

$$\dot{X}(t) = [A + BKB'P^{-1}(t)]X(t) \quad (4.153)$$

$$X(0) = I$$

and $F_2(\delta)$ is obtained by replacing δ for $t - \sigma$ in (4.150). Specifically, if we are interested in the interpolated value of $x(a)$ where a is constant, (4.152) yields

$$\hat{x}^{(\sigma)}(a) = X(a)X^{-1}(t)\hat{x}(t) + \int_a^t X(a)X^{-1}(\delta)F_2(\delta)\hat{x}(\delta)\,d\delta \quad (4.154)$$

where $\hat{x}^{(\sigma)}(a)$ denotes the best estimate of $x(a)$ given the observation $y(\tau)$, $0 \leq \tau \leq a + \sigma, a + \sigma = t$.

Equation (4.154) can be simplified somewhat. Let us differentiate both sides of (4.154) with respect to t. It follows that

$$\frac{d\hat{x}^{(\sigma)}(a)}{dt} = X(a)\left[\frac{d}{dt} X^{-1}(t)\right]\hat{x}(t) + X(a)X^{-1}(t)\frac{d}{dt}\hat{x}(t)$$
$$+ X(a)X^{-1}(t)F_2(t)\hat{x}(t) \quad (4.155)$$

Since $X(t)$ satisfies (4.153) then*

$$\frac{d}{dt} X^{-1}(t) = -X^{-1}(t)[A + BKB'P^{-1}(t)] \quad (4.156)$$

Substituting (4.156) and the expressions for $\frac{d}{dt} x(t)$ from (4.51) and F_2 from (4.145) into (4.155) yields

$$\frac{d\hat{x}^{(\delta)}(a)}{dt} = X(a)X^{-1}(t)P(t)C'[DLD]^{-1}\{y(t) - C\hat{x}(t)\} \quad (4.157)$$

Integrating (4.157) over the interval $[a, t]$ it follows that

$$\hat{x}^{(\sigma)}(a) = \hat{x}(a) + \int_a^t X(a)X^{-1}(s)P(s)C'[DLD']^{-1}\{y(s) - C\hat{x}(s)\}\, ds \quad t \geq a$$
$$(4.158)$$

Clearly, since $\sigma = t - a$, when $t = a$ ($\sigma = 0$) then $\hat{x}^{(\sigma)}(a) = \hat{x}(a)$ as expected. Equation (4.158) is a recursive relationship for generating $\hat{x}^{(\sigma)}(a)$, since as t grows we only need to add a correction term specified by the integral in the right hand side of (4.158) to the previous estimate in order to arrive at the new estimate. This equation is the continuous version of (4.129).

PROBLEMS

1. Given the scalar random sequence $x(k)$ and the scalar observation $y(k)$ described by

$$x(k + 1) = x(k) + u(k)$$
$$y(k) = x(k) + v(k)$$
$$Eu(k) = Ev(k) = 0$$
$$Eu(k_1)u(k_2) = \Delta(k_2 - k_1)$$
$$Ev(k_1)v(k_2) = 2\Delta(k_2 - k_1)$$

* (4.156) is easily obtained by differentiating both sides of the identity $X(t)X^{-1}(t) = I$ with respect to t and utilizing (4.153).

Derive the optimal estimate $x(k + 1)$ given $y(1), \ldots, y(k)$ and specifically the estimate $x(4)$ given $y(1) = 2$, $y(2) = 3$, $y(3) = 1$. Discuss the behavior of the estimator and the covariance matrix $P(k)$ as k approaches infinity.

2. In radar tracking of an object by means of track-while-scan we make observation of the continuous target trajectory at some discrete intervals. However, we generally desire to derive the best estimate of the target trajectory. A one-dimensional example of such process is described by $x(t)$ as a continuous process and $y(kT)$ as the discrete observation such that

$$\dot{x}(t) = -x(t) + u(t)$$
$$y(kT) = x(kT) + v(kT)$$
$$T = 1$$
$$Eu(t) = Ev(k) = 0$$
$$Eu(t_1)u(t_2) = \delta(t_2 - t_1)$$
$$Ev(k_1T)v(k_2T) = \Delta(k_2 - k_1)$$
$$Eu(t_1)v(k_2T) = 0$$

Derive the minimum mean square estimator of $x(t)$ for all t.

3. Still a more practical version of Problem 2 is when we consider that the radar beam width is not zero. Hence the observations are received again at discrete intervals but each observation is spread over some nonzero time interval. In the above problem let the observation be given by

$$y(kT + \tau) = x(kT + \tau) + v(kT + \tau), \qquad k = 0, 1, \ldots, \qquad 0 \leq \tau \leq \tau_0$$
$$T = 1 \qquad \tau_0 = 0.1$$
$$Ev(t) = 0$$
$$Ev(t_1)v(t_2) = \delta(t_2 - t_1)$$

Derive the minimum mean square estimator of $x(t)$ for all t.

4. Show that the estimators derived in examples of Sections 4.4 and 4.7 satisfy the appropriate orthogonality condition.

5. Derive expressions for covariance matrices of estimation errors $\tilde{x}(t + \delta)$ and $\tilde{x}(k + \delta)$, discussed in Section 4.10 and 4.11, in terms of $P(t)$ and $P(k)$, respectively

6. Using the results of Problem 5, discuss how the matrix $P(t_2)$, introduced in Section 4.12, can be evaluated.

SELECTED READINGS

Fagin, S. L., "Recursive Linear Regression Theory," *IEEE International Convention Record*, part I. New York, 1964.

Kalman, R. E., *New Methods and Results in Linear Prediction and Estimation Theory*, Tech. Report 61-I, Research Inst. for Advanced Study, Baltimore, Md., 1961.

Kalman, R. E., "A New Approach to Linear Filtering and Prediction Problems," *ASME Trans. (J. Basic Engineering)*, Vol. 82D, March 1960.

Kalman, R. E., and R. C. Bucy, "New Results in Linear Filtering and Prediction Theory," *ASME Trans. (J. Basic Engineering)*, Vol. 83D, March 1961.

Meditch, J. S., *Orthogonal Projection and Discrete Optimal Linear Smoothing*, Report No. 66-102, Information Processing and Control System Laboratory, Northwestern University, Evanston, Ill., July 1966.

Papoulis, A., *Probability, Random Variables, and Stochastic Processes*, McGraw-Hill, New York, 1965.

Rauch, H. E., "Solutions to the Smoothing Problem," *IEEE Transactions on Automatic Control*, AC-8, 1963.

Swerling, P., "Topics in Generalized Least Squares Signal Estimation," *J. SIAM Applied Mathematics*, Vol. 14, No. 5, September 1966.

Wiener, N., *The Extrapolation, Interpolation, and Smoothing of Stationary Time Series with Engineering Applications*, Wiley, New York, 1949.

chapter 5/
TIME-INVARIANT
LINEAR ESTIMATORS

5.1 I roduction

When the quantity to be estimated and the additive noise in the models of the previous chapter are characterized by wide-sense stationary random processes (so that their autocorrelations exist), and if the interval over which the observation is made is infinite or semi-infinite, the optimal estimator minimizing the average square (or other quadratic form) of the estimation error can be described by a linear constant-coefficient differential (or difference) equation. In certain cases, these constant-coefficient estimators can be obtained by determining the limiting case of the estimators derived in the preceding chapter as the interval of observation approaches infinity. Although in real life the interval of observation can never be infinite, it is often approximated as such, in order to be able to derive a time-invariant filter. This is because linear constant-coefficient estimators (Wiener filters) are often easier to derive and construct than are linear, time-variable estimators (Kalman-type estimators).

In this chapter, the classical procedure for deriving optimal time-invariant filters will be introduced. In comparison with the technique of finding the limiting case of the general linear, time-variable estimator of the preceding chapter, this classical procedure is simpler to apply in many cases and, furthermore, yields a broader class of filters than those which can be described by the "normal form" differential equation (4.37) or difference equation (4.9).

5.2 Continuous Time-Invariant Estimator, Problem Statement

Let $s(t)$ and $n(t)$ be zero-mean wide-sense stationary scalar random processes with known correlation functions given by

$$Es(t_1)s(t_2) = \varphi_{ss}(t_2 - t_1) \tag{5.1}$$

$$En(t_1)n(t_2) = \varphi_{nn}(t_2 - t_1) \tag{5.2}$$

$$Es(t_1)n(t_2) = \varphi_{sn}(t_2 - t_1) \tag{5.3}$$

Consider any time-invariant (but not necessarily causal) linear functional $w(t)$ of the signal $s(t)$ defined by the following infinite convolution integral:

$$w(t) = \int_{-\infty}^{+\infty} g(\lambda)s(t - \lambda)\, d\lambda \tag{5.4}$$

It is desired to determine, for every t, an estimate $\hat{w}(t)$ for $w(t)$ which minimizes

$$E[w(t) - \hat{w}(t)]^2 \tag{5.5}$$

by processing only the "present and past" values of the observed data

$$y(\tau) = s(\tau) + n(\tau); \quad -\infty \le \tau \le t \tag{5.6}$$

Let us further require $\hat{w}(t)$ to be a causal *linear* functional of $y(\tau)$, representable by the semi-infinite convolution integral $\hat{w}(t) = \int_0^\infty h(\tau)y(t - \tau)\, d\tau$. A few examples of the function $g(t)$ in (5.4) are given below.

If $g(t) = \delta(t)$, then $w(t) = s(t)$. This is the filtering problem since we are interested in the estimate of the signal $s(t)$ at time t.

If $g(t) = \delta(t - t_0)$, $t_0 > 0$, then $w(t) = s(t - t_0)$. This is the interpolation problem.

If $g(t) = \delta(t + t_0)$, $t_0 > 0$ then $w(t) = s(t + t_0)$. This yields the prediction problem since we wish to estimate $s(t + t_0)$ given $y(\tau)$; $\tau \le t$.

If $g(\tau) = e^{-\alpha\tau}$, $\tau \ge 0$, and $g(\tau) = 0$ otherwise, then

$$w(t) = \int_0^\infty e^{-\alpha\lambda}s(t - \lambda)\, d\lambda = \int_{-\infty}^t e^{-\alpha(t-\tau)}s(\tau)\, d\tau.$$

5.3 Derivation of Continuous Estimator

In a continuous estimation problem, we are usually given the power spectral density functions of the signal and noise processes, together with some "desired" functional involving a given weighting function $g(\lambda)$, as in (5.4). We wish to determine the optimal *causal* estimator weighting function $h(\lambda)$.

Following the procedure used in Chapter 2, the first step is the determination of the condition which the optimal estimator weighting function $h(\lambda)$, and consequently the estimator itself, should satisfy. The filter is postulated to be linear and causal. In general, it can be represented by the following convolution integral

$$\hat{w}(t) = \int_0^\infty h(\lambda)y(t - \lambda)\,d\lambda \qquad (5.7)$$

The lower bound on the integral is chosen to be zero rather than infinity to indicate that $\hat{w}(t)$ cannot be a function of observations which are not available at time t, namely $y(t - \lambda)$, $\lambda < 0.$* The upper bound is infinite since the entire observation over the interval $-\infty \leq \tau \leq t$ is available to the estimator.

Following steps identical with those in Section 4.6, the principle of orthogonality for the present linear mean-square estimation problem can be derived as

$$E[w(t) - \hat{w}(t)]y(t - \tau) = 0 \qquad 0 \leq \tau \leq +\infty \qquad (5.8)$$

Note that (5.8) forces the estimation error $w(t) - \hat{w}(t)$ to be orthogonal to the entire data waveform $y(t - \tau)$; $0 \leq \tau \leq \infty$. Substituting for $w(t)$ and $\hat{w}(t)$ from (5.4) and (5.7) into (5.8) yields

$$E\left[\int_{-\infty}^{+\infty} g(\lambda)s(t - \lambda)\,d\lambda - \int_0^\infty h(\lambda)y(t - \lambda)\,d\lambda\right]y(t - \tau) = 0 \qquad \tau \geq 0$$

$$(5.9)$$

Taking the expectation inside the integrals results in

$$\int_{-\infty}^{+\infty} g(\lambda)\varphi_{sy}(\tau - \lambda)\,d\lambda - \int_0^\infty h(\lambda)\varphi_{yy}(\tau - \lambda)\,d\lambda = 0 \qquad \tau \geq 0 \quad (5.10)$$

where the terms $\varphi_{sy}(\tau)$ and $\varphi_{yy}(\tau)$ are correlation functions between s, y and y, y respectively. These functions are readily obtained from (5.1)–(5.3) and (5.6).† Equation (5.10) is called the Wiener-Hopf integral equation. Note that the integral equation contains one unknown, namely the weighting function $h(\lambda)$. We must now solve this integral equation for $h(\lambda)$. Since $\varphi_{sy}(\tau)$ and $\varphi_{yy}(\tau)$ are presumed known, and if we are to assume that the weighting functions $g(t)$ and $h(t)$ were given, then the left-hand side of (5.10) could be regarded as defining a function $f(\tau)$ which would satisfy

$$f(\tau) = 0 \qquad \tau \geq 0 \qquad (5.11)$$

We shall now employ Fourier transforms of the various functions thus far introduced in order to continue the derivations. Let $F(j\omega)$ be the Fourier

* Actually $h(\lambda)$ is chosen to be zero for $\lambda < 0$.
† $\varphi_{sy}(\tau) = \varphi_{ss}(\tau) + \varphi_{sn}(\tau)$; $\varphi_{yy}(\tau) = \varphi_{ss}(\tau) + \varphi_{nn}(\tau) + 2\varphi_{sn}(\tau)$.

transform* of $f(\tau)$. Since $f(\tau)$ is, by (5.11), equal to zero for $\tau \geq 0$ it follows that the function $F(s)$, where s is a complex variable, is analytic in the left half of the complex s-plane.† Taking the Fourier transform of (5.10), noticing that the integrals in (5.10) are convolutions, and substituting s for $j\omega$ yields

$$G(s)\phi_{sy}(s) - H(s)\phi_{yy}(s) = F(s) \tag{5.12}$$

where

$$\phi(s) \triangleq \phi(j\omega) = \frac{1}{2\pi} \int_{-\infty}^{+\infty} \varphi(\tau)e^{-j\omega\tau}\,d\tau$$

and $F(s)$ is analytic in the left half plane. By making use of the condition on $F(s)$, equation (5.12) will be solved in the following for $H(s)$ without the actual knowledge of $F(s)$.

Let us introduce the function $\phi_{yy}^+(s)$ which is analytic in right half plane and the function $\phi_{yy}^-(s)$ which is analytic in the left half plane and is derived from $\phi_{yy}(s)$ by

$$\phi_{yy}^-(s) = \phi_{yy}^+(-s) \tag{5.13}$$

$$\phi_{yy}(s) = \phi_{yy}^-(s)\phi_{yy}^+(s) \tag{5.14}$$

The operation yielding ϕ_{yy}^- and ϕ_{yy}^+ from ϕ_{yy} is referred to as spectral factorization.

Let us now expand the expression $\dfrac{G(s)\phi_{sy}(s)}{\phi_{yy}^-(s)}$ by partial fractions into a sum of terms $F_1(s)$ and $F_2(s)$, where F_1 and F_2 are required to be analytic in right half plane and left half plane respectively to get

$$\frac{G(s)\phi_{sy}(s)}{\phi_{yy}^-(s)} \triangleq F_1(s) + F_2(s) \tag{5.15}$$

Using (5.14) and (5.15), equation (5.12) can be written as

$$F_1(s) - H(s)\phi_{yy}^+(s) = \frac{F(s)}{\phi_{yy}^-(s)} - F_2(s) \tag{5.16}$$

Because of the way we have constructed (5.15), all of the terms in the left hand side of (5.16) are analytic in the right half plane and all terms in the

---

* The Fourier transform of $f(\tau)$ exists if each term in (5.10) has a Fourier integral. This imposes the following conditions: (a) $\int_{-\infty}^{\infty} |g(\lambda)|\,d\lambda < \infty$, which imposes a restriction on $g(t)$; (b) $\int_{-\infty}^{\infty} |\varphi(\tau)|\,d\tau < \infty$ is usually satisfied in practice, for example, if $\varphi(\tau)$ is bounded, differentiable in τ and $\lim_{\tau \to \infty} \varphi(\tau) = 0$; (c) $\int_{-\infty}^{+\infty} |h(\lambda)|\,d\lambda < \infty$. This indicates that we are limiting the search for optimal systems to stable systems since we can identify $h(t)$ as the impulse response of a linear constant-coefficient system. This is actually not a restriction, since it can easily be shown that an unstable system cannot possibly be optimal.

† In fact $F(s)$ is the Laplace transform of $f(\tau)$.

right hand side are analytic in the left half plane. This is possible only if both sides of (5.16) are equal to a constant, which we take as zero. Consequently, a solution to equation (5.16) is

$$F_1(s) - H(s)\phi_{yy}^+(s) = 0 \tag{5.17}$$

or

$$H(j\omega) = H(s) = \frac{F_1(s)}{\phi_{yy}^+(s)} \tag{5.18}$$

This is the transfer function for the optimal causal filter, i.e., the Fourier transform of $h(t)$.

5.4 An Alternative Derivation

We derived in Section 5.3 that the necessary condition for $h(\lambda)$ to be the weighting function of the optimal filter is that $h(\lambda)$ satisfy equation (5.10). This relationship is to be satisfied for all non-negative values of τ. If, for the

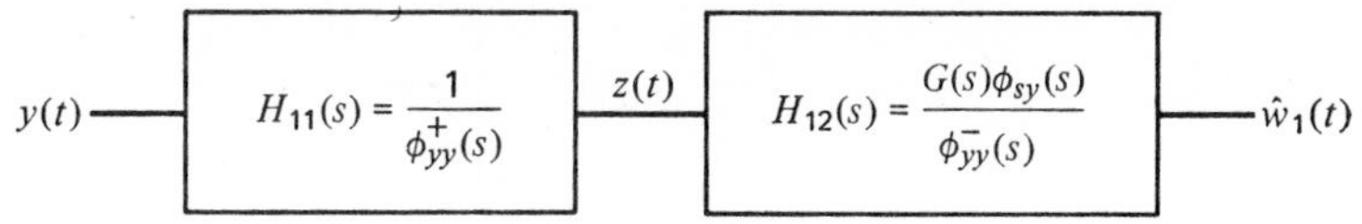

figure 5.1

time being, we assumed that the observation $y(\tau)$ over the interval $-\infty \leq \tau \leq +\infty$ is available for estimating $w(t)$, then this restriction does not exist, and a simple expression for $H(s)$ can be obtained by changing limits of integration in (5.10) from $(0, \infty)$ to $(-\infty, +\infty)$ and then taking the Fourier transform of (5.10) to obtain

$$H_1(s) = \frac{G(s)\phi_{sy}(s)}{\phi_{yy}(s)} \tag{5.19}$$

The subscript on H indicates that (5.10) is satisfied for all τ rather than for $\tau \geq 0$. Using "spectral factorization," as in (5.14), equation (5.19) can be decomposed and written as

$$H_1(s) = \frac{1}{\phi_{yy}^+(s)} \frac{G(s)\phi_{sy}(s)}{\phi_{yy}^-(s)} \tag{5.20}$$

The transfer functions appearing in (5.20) can be rearranged to obtain a convenient form for the estimator, as represented by the block diagram in Figure 5.1.

It is interesting to determine the spectral density of the random process $z(t)$ at the intermediate point indicated in Figure 5.1. Using the expression for the power spectral density function of the output of a linear filter $\dfrac{1}{\phi_{yy}^+(s)}$ in response to an input* with power spectral density $\phi_{yy}(j\omega)$, we obtain:

$$\phi_{zz}(j\omega) = \phi_{yy}(j\omega) \left| \frac{1}{\phi_{yy}^+(j\omega)} \right|^2 \triangleq \phi_{yy}(j\omega) \frac{1}{\phi_{yy}^-(j\omega)} \cdot \frac{1}{\phi_{yy}^+(j\omega)} = 1 \quad (5.21)$$

Notice that the magnitude-squared function $\left| \dfrac{1}{\phi_{yy}^+(j\omega)} \right|^2$ involves, by definition, $\phi_{yy}^-(j\omega)$ and its complex conjugate $\phi_{yy}^-(j\omega)$. Therefore, $z(t)$ is a random process with an autocorrelation function given by $\delta(\tau)$ (equivalent to unity power spectral density). Now if we re-introduce the restriction that $y(t)$

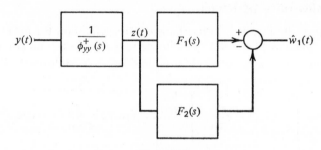

figure 5.2

is observed only over the interval $-\infty \leq \tau \leq t$, then $z(t)$ will be given only over this interval. Since $z(t)$ is an uncorrelated process $\big($i.e., $R_{zz}(\tau) = \delta(\tau)\big)$, no physically realizable system $\big($such as $H_{12}(s)$ operating on $z(t)\big)$ can produce an output at t which is a function of a future input at $\tau > t$. Consequently, the inverse transform of $H_{12}(s)$, namely $h_{12}(t)$, should be zero for $t \leq 0$. Unfortunately, it can easily be shown that this condition is not satisfied. We can, however, fix up this difficulty.

Using (5.15) Figure 5.1 can be represented in equivalent form as shown in Figure 5.2

The inverse Fourier transforms of $F_1(j\omega)$ and $F_2(j\omega)$ denoted, respectively by $f_1(t)$ and $f_2(t)$, have the properties

$$f_1(t) = 0 \qquad t < 0$$
$$f_2(t) = 0 \qquad t > 0$$

because of the restriction that F_1 was analytic in the left half plane and F_2 was analytic in the right half plane.

Consequently, the function $F_1(s)$ operates on the past data $\big(y(\tau), \tau < t\big)$ and $F_2(s)$ operates on the future data $\big(y(\tau), \tau > t\big)$. Since the future data

* *See* Section 1.41, Chapter 1.

are not available, the term $F_2(s)$ cannot be physically constructed, and hence it will be dropped. This yields the optimal estimator given by equation (5.18). The above procedure is referred to as the Bode-Shannon method. This method, although it appears to be motivated by heuristic physical considerations, does in fact develop an estimator transfer function which is the rigorous solution to the causal, time-invariant estimator problem.

5.5 Estimation Error

As usual, the estimation error is defined by $E[w(t) - \hat{w}(t)]^2$. An expression for the estimation error is obtained by determining the spectral density of $w(t) - \hat{w}(t)$ and integrating the result* over $-\infty < \omega < +\infty$

$$E[w(t) - \hat{w}(t)]^2 = \frac{1}{2\pi} \int_{-\infty}^{+\infty} [|G(j\omega)|^2 \phi_{ss}(j\omega) + |H(j\omega)|^2 \phi_{yy}(j\omega)$$
$$- 2G(j\omega)H(-j\omega)\phi_{sy}(j\omega)] \, d\omega \quad (5.22)$$

where $H(j\omega)$ is given by (5.18). Equation (5.22) can be used to evaluate how good the optimal estimator is, and furthermore, can serve as a standard of comparison for evaluation of performance of any sub-optimal filter which may be used in a given application because of simplicity in construction, reliability, or other practical considerations.

5.6 Examples

▶ **example 1**

Let the signal $s(t)$ have zero-mean and autocorrelation function given by

$$Es(t_1)s(t_2) = \tfrac{3}{2}e^{-|t_2-t_1|} \quad (5.23)$$

We would like to derive the optimal estimate $\hat{s}(t)$ of the signal $s(t)$ by processing the observed data $y(\tau) = s(\tau) + n(\tau)$, $-\infty \leq \tau \leq t$ where $n(t)$ is additive noise with the properties

$$En(t) = 0$$
$$En(t_1)n(t_2) = \delta(t_2 - t_1) \quad (5.24)$$
$$En(t_1)s(t_2) = 0$$

From (5.23) and (5.24)

$$\phi_{sy}(j\omega) = \phi_{ss}(j\omega) = \frac{3}{1 + \omega^2}$$

$$\phi_{yy}(j\omega) = \frac{4 + \omega^2}{1 + \omega^2}$$

* By applying the Parseval's theorem equation (1.205), Chapter 1.

Hence

$$\phi_{yy}^+(j\omega) = \frac{2 + j\omega}{1 + j\omega}$$

Since $G(j\omega) = 1$ for this problem

$$\frac{G(j\omega)\phi_{sy}(j\omega)}{\phi_{yy}^-(j\omega)} = \frac{3}{(1 + j\omega)(2 - j\omega)} = \frac{1}{1 + j\omega} + \frac{1}{2 - j\omega}$$

Hence

$$F_1(s) = \frac{1}{1 + s}$$

or finally from (5.18)

$$H(s) = \frac{1}{2 + s}$$

which defines a filter with one-sided Laplace transfer function $\dfrac{1}{s + 2}$

$$\underrightarrow{y(t)} \boxed{\frac{1}{s + 2}} \underrightarrow{\hat{s}(t)}$$

This filter can also be described by the differential equation

$$\dot{\hat{s}}(t) = -2\hat{s}(t) + y(t)$$

This result agrees with the result of Example 1 in Section 4.9 in the limiting ▶ case of infinite past observation time.

▶ **example 2**
This example is given to demonstrate that an estimator obtained by the "classical" method of this chapter may not belong to the class of estimation which can be characterized by a differential equation of the "normal" form (4.37). Let $s(t)$ and $n(t)$ be zero-mean random processes with correlation functions given by

$$\varphi_{ss}(\tau) = e^{-|\tau|}$$

$$\varphi_{sn}(\tau) = 0$$

$$\varphi_{nn}(\tau) = e^{-2|\tau|}$$

It is desired to construct the optimal estimate $\hat{s}(t)$ of $s(t)$ from data $y(\lambda) = s(\lambda) + n(\lambda)$; $-\infty \leq \lambda \leq t$. From the above spectral specifications, we have

$$\phi_{sy}(j\omega) = \phi_{ss}(j\omega) = \frac{2}{1 + \omega^2}$$

$$\phi_{yy}(j\omega) = \frac{6(2 + \omega^2)}{(1 + \omega^2)(4 + \omega^2)}$$

Hence

$$\phi_{yy}^+(j\omega) = \frac{\sqrt{6}\,(\sqrt{2} + j\omega)}{(1 + j\omega)(2 + j\omega)}$$

and

$$F_1(j\omega) = \frac{\sqrt{6}}{1 + \sqrt{2}} \cdot \frac{1}{1 + j\omega}$$

Finally

$$H(j\omega) = \frac{1}{1 + \sqrt{2}} \cdot \frac{2 + j\omega}{\sqrt{2} + j\omega}$$

which gives the optimum filter as the one-sided Laplace transfer function
$$\frac{1}{1 + \sqrt{2}} \cdot \frac{2 + s}{\sqrt{2} + s} \cdot$$

$$y(t) \quad \boxed{\dfrac{1}{1 + \sqrt{2}} \cdot \dfrac{2 + s}{\sqrt{2} + s}} \quad \hat{s}(t)$$

Clearly this filter cannot be represented by (4.37), since we cannot derive a differential equation for $\hat{s}(t)$ with only $y(t)$ (but not $\dot{y}(t)$) as the input. However, the procedure of the preceding chapter (with some modifications) ▶ can yield the solution to the present problem (see Chapter 6).

▶ **example 3**
Using the data of Example 1, let us determine the estimator yielding $\hat{s}(t + t_0)$, $t_0 > 0$. In other words, we should like to estimate a future value of signal based on the observed data $y(\tau)$; $-\infty \leq \tau \leq t$. This can be done by letting $g(\lambda)$ in (5.4) be equal to $\delta(\lambda + t_0)$ or $G(s) = e^{t_0 s}$. Again, we have

$$\phi_{yy}^+(j\omega) = \frac{2 + j\omega}{1 + j\omega}$$

Due to the term $e^{t_0 s}$ the decomposition suggested by (5.15) cannot be obtained easily. Instead, we may derive the portion of the inverse transform of the function $\dfrac{G(s)\phi_{sy}(s)}{\phi_{yy}^-(s)}$ corresponding to $t > 0$ and use its transform, which represents $F_2(s)$. Hence

$$F_2(s) = \frac{e^{-t_0}}{1 + s}$$

The estimator is then given by

$$H(s) = \frac{e^{-t_0}}{2 + s}$$

It is seen that when $t_0 = 0$, the result agrees with that of Example 1. Furthermore, when t_0 approaches infinity the function $H(\omega)$ approaches zero. This is intuitively plausible, since the value of the signal at an infinite time in the future tends to become uncorrelated with its present and past value of the observation (i.e., $\lim_{\tau \to \infty} \varphi_{ss}(\tau) = 0$).

The problem is somewhat similar when we would like to estimate $s(t - t_0)$, $t_0 > 0$, given $y(\tau)$, $-\infty \leq \tau \leq t$. Here, (5.15) takes the form

$$\frac{G(s)\phi_{sy}(s)}{\phi_{yy}^-(s)} = \frac{3}{e^{st_0}(1 + s)(2 - s)} \triangleq F(s)$$

The decomposition of this function, denoted by $F(s)$, according to equation (5.15) into functions $F_1(s)$ and $F_2(s)$ is again made complicated by the existence of the term e^{st_0} in the denominator.* It is easier to perform this

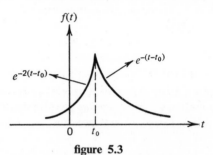

figure 5.3

operation by using the inverse transform of $F(s)$ shown graphically in Figure 5.3. Then $F_1(s)$ has an inverse transform $f_1(t)$ given by

$$f_1(t) = f(t) \qquad t \geq 0$$

▶
$$f_1(t) = 0 \qquad t < 0$$

5.7 Discrete Estimator, Problem Statement

Let $s(k)$ and $n(k)$ be zero-mean wide-sense stationary random sequences with known correlation functions given by

$$Es(k_1)s(k_2) = \varphi_{ss}(k_2 - k_1) \tag{5.25}$$

$$En(k_1)n(k_2) = \varphi_{nn}(k_2 - k_1) \tag{5.26}$$

$$Es(k_1)n(k_2) = \varphi_{sn}(k_2 - k_1) \tag{5.27}$$

Consider any time-invariant linear functional $w(k)$ of the signal sequence $s(-\infty), \ldots, s(k - 1), s(k), s(k + 1), \ldots, s(+\infty)$ which can be represented

* e^{st_0} has an essential singularity at infinity.

by the infinite "convolution sum"

$$w(k) = \sum_{i=-\infty}^{+\infty} g(i)s(k - i) \qquad (5.28)$$

It is desired to determine, for every k, an estimate $\hat{w}(k)$ for $w(k)$ which minimizes

$$E[w(k) - \hat{w}(k)]^2 \qquad (5.29)$$

by processing only the "present and past" values of the observed data sequence (i.e., for $-\infty \leq i \leq k$)

$$y(i) = s(i) + n(i), \qquad (-\infty \leq i \leq k) \qquad (5.30)$$

Let us require $\hat{w}(k)$ to be a linear functional of these $y(i)$, representable by a semi-infinite "convolution sum" of the form $\hat{w}(k) = \sum_{i=0}^{\infty} h(i)y(k - i)$. The discrete weighting sequences $g(i)$, $-\infty \leq i \leq +\infty$, and $h(i)$, $0 \leq i \leq \infty$, are similar to the functions $g(t)$, $-\infty \leq t \leq +\infty$, introduced in Section 5.2, and can be given identical interpretations.

5.8 Derivation of Discrete Filter

The procedure to be followed here is very similar to that of Section 5.3. The filter is postulated to be linear and time-invariant, and hence can in general be presented by the following convolution sum:

$$\hat{w}(k) = \sum_{i=0}^{\infty} h(i)y(k - i) \qquad (5.31)$$

As in Section 5.2, the lower bound on the summation is set equal to zero to impose the physical realizability condition that the estimate $\hat{w}(k)$ cannot be a function of $y(j)$, $j > k$ (i.e., the "future" observations which have not yet been received at k). The upper bound is infinity, since the entire past and present observation sequence $y(i)$, $-\infty \leq i \leq k$, is postulated to be at the disposal of the estimator.

To obtain the optimal linear least-square estimator, we must use the necessary and sufficient condition that the estimator error $w(k) - \hat{w}(k)$, for *every* k, be orthogonal to *all* the past and present data. Thus, the principle of orthogonality requires

$$E[w(k) - \hat{w}(k)]y(k - j) = 0 \qquad 0 \leq j \leq +\infty \qquad (5.32)$$

Substituting for $w(k)$ and $\hat{w}(k)$ from (5.28) and (5.31) into (5.32) yields

$$E\left[\sum_{-\infty}^{+\infty} g(i)s(k - i) - \sum_{0}^{\infty} h(i)y(k - i)\right]y(k - j) = 0 \qquad j \geq 0 \qquad (5.33)$$

Taking the expectation inside the summations, and using the fact that the processes $s(k)$, $n(k)$, and consequently $y(k)$ are (wide-sense) stationary, we obtain

$$\sum_{i=-\infty}^{+\infty} g(i)\varphi_{sy}(j-i) - \sum_{i=0}^{\infty} h(i)\varphi_{yy}(j-i) = 0 \qquad j \geq 0 \qquad (5.34)$$

where φ_{sy} and φ_{yy} are correlation functions between s and y, and y and y, respectively, which can be derived from (5.25) through (5.27). Since φ_{sy} and φ_{yy} are presumed known, and if we were to assume that the weighting sequences $g(i)$ and $h(i)$ were given, then the left hand side of (5.34) could be regarded as defining a function $f(j)$ which would satisfy

$$f(j) = 0 \qquad j \geq 0 \qquad (5.35)$$

We shall now employ discrete, two-sided Laplace transforms of the various sequences thus far introduced, in order to continue our derivations. The two-sided Laplace transform of the sequence $f(j)$ is defined as

$$F^*(s) \triangleq \sum_{-\infty}^{+\infty} f(j)e^{-sj} \qquad (5.36)$$

Since $f(j)$ is, by (5.35), equal to zero for $j \geq 0$, it follows that its transform $F^*(s)$ is analytic in the right half of the complex s-plane. We may simplify the notation of (5.36) somewhat by introducing the change of variables $z = e^s$, which has become standard in work with discrete sequences. Using this change of variables, (5.36) becomes

$$F(z) \triangleq \sum_{-\infty}^{+\infty} f(j)z^{-j} \qquad (5.36a)$$

The transform $F(z)$ is called, simply, the "z-transform" of the discrete sequence $f(j)$. Since $F^*(s)$ was analytic in the right half of the s-plane, $F(z)$ is then analytic inside the unit circle in the complex z-plane. Let us define the following z-transforms, as in (5.36a)

$$G(z) \triangleq \sum_{-\infty}^{+\infty} g(j)z^{-j} \qquad (5.37)$$

$$\phi(z) \triangleq \sum_{-\infty}^{+\infty} \varphi(k)z^{-k} \qquad (5.38)$$

$$H(z) \triangleq \sum_{0}^{+\infty} h(j)z^{-j} \qquad (5.39)$$

Furthermore, let us define the following functions

$$\phi_{yy}^-(z) \triangleq \phi_{yy}^+(z^{-1}) \qquad (5.40)$$

$$\phi_{yy}(z) \triangleq \phi_{yy}^-(z)\phi_{yy}^+(z) \qquad (5.41)$$

and

$$\frac{G(z)\phi_{sy}(z)}{\phi_{yy}^-(z)} = F_1(z) + F_2(z) \qquad (5.42)$$

where F_1 and F_2 are analytic inside and outside of a unit circle in z-plane respectively. By finding the z-transform of (5.34) and making use of the identities (5.40), (5.41), and (5.42), the following expression for the optimal filter is derived

$$H(z) = \frac{F_1(z)}{\phi_{yy}^+(z)} \qquad (5.43)$$

The procedure employed here is similar to that leading to equation (5.18) of the continuous case.

▶ example

Let the statistical properties of the random sequences $s(k)$ and $n(k)$ be:

$$Es(k) = En(k) = 0$$
$$Es(k_1)s(k_2) = e^{-|k_2-k_1|}$$
$$En(k_1)n(k_2) = \Delta(k_2 - k_1)$$
$$Es(k_1)n(k_2) = 0$$

It is desired to derive the estimate $\hat{s}(k)$ of $s(k)$ which minimizes $E[s(k) - \hat{s}(k)]^2$. This corresponds to using $g(i) = \Delta(i)$ in (5.28) so that $w(k)$ is actually equal to $s(k)$, and $G(z) = 1$. We have

$$G(z) = 1$$
$$\phi_{ss}(z) = \sum_{-\infty}^{+\infty} e^{-|k|} z^{-k} = \frac{1 - e^{-2}}{(1 - e^{-1}z^{-1})(1 - e^{-1}z)}$$
$$\phi_{nn}(z) = 1$$

Hence

$$\phi_{yy}(z) = \frac{2 - e^{-1}z - e^{-1}z^{-1}}{(1 - e^{-1}z^{-1})(1 - e^{-1}z)}$$
$$\phi_{sy}(z) = \phi_{ss}(z)$$

From (5.40) and (5.41) we get

$$\phi_{yy}^+(z) = \sqrt{1 - \sqrt{1 - e^{-2}}} \left[\frac{z^{-1} - (e + \sqrt{e^2 - 1})}{1 - e^{-1}z^{-1}} \right]$$

$$F_1(z) = \frac{1 - e^{-2}}{\sqrt{e^{-1} - (e + \sqrt{e^2 - 1})}\sqrt{1 - \sqrt{1 - e^2}}} \frac{1}{1 - e^{-1}z^{-1}}$$

Finally, from (5.43), the transfer functions of the optimal estimator (which can be called a "filter") is given by the z-transform

$$H(z) = \frac{\alpha}{1 - \beta z^{-1}}$$

where

$$\alpha = \frac{(e^{-2} - 1)(e - \sqrt{e^2 - 1})}{\sqrt{e^{-1} - (e + \sqrt{e^2 - 1})}}$$

$$\beta = e - \sqrt{e^2 - 1}$$

The corresponding filter weighting sequence $h(k)$ is readily obtained as

$$h(k) = \alpha \beta^k$$

The difference equation representing the optimal filter is given by

▶ $$\hat{s}(k) = \beta \hat{s}(k - 1) + \alpha y(k)$$

PROBLEMS

1. Derive the orthogonality conditions given by (5.8) and (5.32).

2. Let the signal $s(t)$ and noise $n(t)$ have zero-means and the following auto-correlations

$$\varphi_{ss}(\tau) = \delta(\tau)$$
$$\varphi_{nn}(\tau) = e^{-\alpha|\tau|}, \qquad \alpha > 0$$
$$\varphi_{sn}(\tau) = 0$$

We observe $y(\tau) = s(\tau) + n(\tau)$; $-\infty \leq \tau < t$. Derive a physically realizable filter producing the least-square estimate $\hat{s}(t)$ of $s(t)$. Discuss the action of the filter as α is increased.

3. Let the signal $s(t)$ and noise $n(t)$ have zero-means and autocorrelation functions given by

$$\varphi_{nn}(\tau) = \delta(\tau)$$
$$\varphi_{ns}(\tau) = 0$$
$$\varphi_{ss}(\tau) = 3/2e^{-|\tau|}$$

The observation is $y(\tau) = s(\tau) + n(\tau)$, $-\infty \leq \tau \leq t$.

(a) Derive an estimator for $\int_{-\infty}^{t} e^{-\alpha(t-\tau)} s(\tau) \, d\tau$. Discuss the action of the estimator as α approaches zero ($\alpha > 0$).

(b) Derive an estimator for $\frac{1}{2}[s(t) + s(t + 2t_0)]$, $t_0 > 0$.

(c) Derive an estimator for $\int_{-\infty}^{t+T} e^{-\alpha(T+t-\tau)} s(\tau) \, d\tau$.

4. Let $s(k)$ and $n(k)$ have the following properties

$$Es(k) = En(k) = 0$$
$$Es(k_1)s(k_2) = e^{-|k_2-k_1|}$$
$$En(k_1)n(k_2) = \Delta(k_2 - k_1)$$
$$Es(k_1)s(k_2) = 0$$

We observe $y(i) = 2s(i) + 3n(i)$, $-\infty \le i \le k$.
Derive estimates for the following
(a) $s(k)$
(b) $s(k + 1)$
(c) $s(k) + s(k + 1)$
(d) $n(k)$

5. Derive the necessary modifications in various estimators (discrete and continuous) if the signal, or noise, or both signal and noise have constant known nonzero mean values.

SELECTED READINGS

Bode, H. W. and C. E. Shannon, "A Simplified Derivation of Linear Least-Square Smoothing and Prediction Theory," *IRE Proc.*, **38**, 1950.

Davenport, W. B. and W. L. Root, *An Introduction to the Theory of Random Signals and Noise*, McGraw-Hill, New York, 1958.

Lanning, J. H., Jr. and R. H. Battin, *Random Processes in Automatic Control*, McGraw-Hill, New York, 1956.

Truxal, J. G., *Control System Synthesis*, McGraw-Hill, New York, 1955.

Wainstein, L. A. and V. D. Zubakov, *Extraction of Signals from Noise*, translation from Russian, Prentice-Hall, Englewood Cliffs, N.J., 1962.

Wiener, N., *Extrapolation, Interpolation, and Smoothing of Stationary Time Series*, MIT Technology Press and Wiley, New York, 1950.

chapter 6 / ADDITIONAL TOPICS ON LINEAR ESTIMATORS

6.1 Introduction

All the models of random processes considered in Chapter 4 contained observations with additive white noise terms. Clearly this is a restriction, since it is possible for the observation noise to be colored (i.e., nonwhite), or there may not be any observation noise at all. These cases are considered in Sections 6.2 and 6.3 where it is shown that the models of these random processes can be manipulated to assume the form considered in Chapter 4. Section 6.4 considers the effect of inaccuracies in *a-priori* model parameters on performance of the optimal estimator and a measure of reliability of an estimate obtained under this condition. Section 6.5 introduces a "fixed" memory approximation to the optimal estimators derived in Chapter 4.

6.2 Measurements Containing Colored Noise

Let the random process be given by

$$\dot{x}(t) = Ax(t) + Bu(t)$$
$$y(t) = Cx(t) + Dv(t) \tag{6.1}$$

where the parameters are defined as in (4.27)–(4.31) except for the zero-mean, colored measurement noise v whose covariance function is given by

$$Ev(t_1)v'(t_2) = L(t_2, t_1) \tag{6.2}$$

In many cases it is possible to consider the colored noise $v(t)$ as the output of a linear system with a white noise input. If so, equation (6.1) can be augmented to include the differential equation generating the observation noise. A similar procedure can be used in the random sequence models. This procedure is best illustrated by means of examples.

▶ **example 1**

Let $x_1(t)$ be a scalar random process given by

$$\dot{x}_1(t) = -2x_1(t) + u_1(t)$$

where $u_1(t)$ is white noise with known covariance function. Let the observed signal be

$$y(t) = x_1(t) + v(t)$$

where $v(t)$ is a random process with zero-mean and covariance function given by

$$Ev(t_1)v(t_2) = e^{-|t_2-t_1|}$$

The colored noise term $v(t)$ can be equivalently thought of as the output of the following linear constant-coefficient differential equation

$$\dot{x}_2(t) = -x_2(t) + u_2(t)$$
$$Eu_2(t_1)u_2(t_2) = 2\,\delta(t_2 - t_1)$$
$$v(t) = x_2(t)$$

Hence the augmented model of the random process can be given as

$$\dot{x}_1(t) = -2x_2(t) + u_1(t)$$
$$\dot{x}_2(t) = -x_2(t) + u_2(t) \tag{6.3}$$
$$y(t) = x_1(t) + x_2(t)$$

where both $u_1(t)$ and $u_2(t)$ represent independent zero-mean white noise processes.

▶ **example 2**

Let a scalar random sequence $x_1(k)$ be given by

$$x_1(k + 1) = \tfrac{1}{2}x_1(k) + u_1(k)$$

and the observation by

$$y(k) = x_1(k) + v(k)$$

where $u_1(k)$ is white noise sequence and $v(k)$ is a zero-mean colored random sequence with covariance function

$$Ev(k_1)v(k_2) = e^{-|k_2-k_1|}$$

Similar to the preceding example, this system can be modeled by

$$x_1(k + 1) = \tfrac{1}{2}x_1(k) + u_1(k)$$
$$x_2(k + 1) = e^{-1}x_2(k) + u_2(k) \qquad (6.4)$$
$$Eu_2(k_1)u_2(k_2) = (1 - e^{-2})\,\Delta(k_2 - k_1)$$
$$y(k) = x_1(k) + x_2(k)$$

where both $u_1(k)$ and $u_2(k)$ represent independent zero-mean white noise
▶ sequences.

With application of the procedure introduced in the foregoing Examples
1 and 2, the random processes in most practical situations involving additive
colored noise in the observed quantities can be modeled with only white noise
terms acting as inputs to the differential (or difference) equations.

In the case of a discrete process, the procedure in Example 2 is all that is
necessary in order to be able to derive the optimal estimator of $x(k + 1)$
given $y(i)$, $0 \le i \le k$ (equations (4.24)–(4.26)). The fact that the observation
$y(k)$ may contain no noise at all (as in 6.4) does not cause any difficulty,
since we can let $L = 0$ in (4.25). To apply the procedure in the continuous
case, however, it is necessary that the covariance matrix of the observation
noise $(DLD'$ in (4.50)–(4.51)) be nonsingular. The following section is
devoted to the case where this condition is *not* satisfied (as in Example 1
above).

6.3   Partially or Completely Noise-Free Measurements

Consider the continuous estimation model given by equations (4.27)–(4.31),
and let DLD' be a singular matrix. The physical interpretation of this
condition is that some linear combinations of components of the observed
vector $y(t)$ contain no white noise term. When such is the case, it follows
from (4.28) that some linear combinations of components of the signal vector
$x(t)$ can be observed exactly. In addition, the successive derivatives of the
noise-free observation* elements, or their linear combinations as long as they
remain noise-free, contain the same information, i.e., the exact measure of
some linear combinations of the elements of $x(t)$.

Consequently, the implication of singularity of the matrix DLD' is that it is
not necessary to estimate all of the components of $x(t)$ since a number of
linear relationships among these components are known. The logical
procedure is then to make use of this additional information. The effect is

---

* learly the estimation problem considered is not changed by considering $y(t)$ and, say,
$\dot{y}(t)$ (if it exists) to be made available to the estimator. In other words, we are implying
that the estimator has the ability to differentiate the input data, if that is useful.

that the order of the system $\left(\text{size of vector } x(t)\right)$ will be reduced by the number of existing linear relationships. What is then left is a problem where the corresponding observation covariance function is nonsingular, whereupon the problem can be solved by the method of Chapter 4. The nature of the steps of the procedure are outlined below in summary.

1. Determine those linear combinations of components of $y(t)$ which do not contain noise.
2. Determine the successive derivatives of each of the above linear functions until a noise term appears.
3. Identify all of the above noise-free linear combinations of components of $y(t)$ and their noise-free derivatives as known (exact) linear combinations of components of $x(t)$.
4. Eliminate the information obtained in Step 3 from the estimation problem.
5. Solve the modified problem as discussed in Chapter 4.

The implementation of the above steps of the procedure is simple and is best illustrated by the following detailed examples.

▶ **example 1**

Let two scalar random processes $x_1(t)$ and $x_2(t)$ (e.g., signal and noise, respectively) have zero-mean and autocorrelation functions given by

$$Ex_1(t_1)x_1(t_2) = e^{-|t_2-t_1|} \tag{6.5}$$

$$Ex_2(t_1)x_2(t_2) = e^{-2|t_2-t_1|} \tag{6.6}$$

$$Ex_1(t_1)x_2(t_2) = 0 \tag{6.7}$$

It is desired to estimate $x_1(t)$ and $x_2(t)$ given the observation

$$y(\tau) = x_1(\tau) + x_2(\tau); \quad 0 \le \tau \le t \tag{6.8}$$

The random processes $x_1(t)$ and $x_2(t)$ can be modeled by the following equations

$$\dot{x}_1(t) = -x_1(t) + u_1(t) \tag{6.9}$$

$$\dot{x}_2(t) = -2x_2(t) + u_2(t) \tag{6.10}$$

$$Eu_1(t_1)u_1(t_2) = 2\,\delta(t_2 - t_1) \tag{6.11}$$

$$Eu_2(t_1)u_2(t_2) = 4\,\delta(t_2 - t_1) \tag{6.12}$$

$$Eu_1(t_1)u_2(t_2) = 0$$

The observation $y(t)$ given by (6.8) contains no noise term. We allow the possibility that the estimator obtain and make use of the derivative of $y(t)$. Differentiating both sides of (6.8) yields

$$\dot{y}(t) = \dot{x}_1(t) + \dot{x}_2(t) \tag{6.13}$$

Substituting from (6.9) and (6.10) yields

$$\dot{y}(t) = -x_1(t) - 2x_2(t) + u_1(t) + u_2(t) \tag{6.14}$$

which clearly contains the white noise component $u_1(t) + u_2(t)$. Therefore, the only known relationship between $x_1(t)$ and $x_2(t)$ is given by (6.8). In other words, we cannot use the instantaneous value of the derivative of the observation to obtain an exact relation between $x_1(t)$ and $x_2(t)$.

The next step is to choose a nonsingular transformation M to define a new variable $z(t)$ given by

$$z(t) = Mx(t) \tag{6.15}$$

where $x(t)$ is the vector $\begin{bmatrix} x_1(t) \\ x_2(t) \end{bmatrix}$. M should be so chosen that when the change of variable is introduced into (6.9) and (6.10), the known linear relationships (for this problem only one) are explicitly revealed.* One candidate for M is the matrix

$$M = \begin{bmatrix} 1 & 1 \\ 0 & 1 \end{bmatrix}$$

In Section 4.6 it was shown that the estimate $x(t)$ minimizing the cost function $E[x(t) - \hat{x}(t)]'Q[x(t) - \hat{x}(t)]$ is independent of matrix Q as long as Q is positive definite. Using the change of variable indicated above has the effect of substituting the matrix $M^{-1\prime}QM^{-1}$ for Q in the cost function. Since $M^{-1\prime}QM^{-1}$ is positive definite when M is nonsingular, the change of variable from $x(t)$ to $z(t)$ does not alter the estimation problem.

Applying the above change of variable to (6.9) and (6.10) results in

$$\dot{z}_1(t) = -z_1(t) - z_2(t) + u_1(t) + u_2(t) \tag{6.16}$$

$$\dot{z}_2(t) = -2z_2(t) + u_2(t) \tag{6.17}$$

$$z_1(t) = x_1(t) + x_2(t) \tag{6.18}$$

$$z_2(t) = x_2(t) \tag{6.19}$$

Therefore $z_1(t)$ is known (it is equal to $y(t)$, which is actually observed) and it remains to estimate $z_2(t)$. Note that $z_1(t) = x_1(t) + x_2(t)$ is indeed a linear combination of components of $x(t)$ which contain no additive white noise.

Equations (6.8) and (6.14) now take the form

$$y(t) = z_1(t) \tag{6.20}$$

$$\dot{y}(t) = -z_1(t) - z_2(t) + u_1(t) + u_2(t) \tag{6.21}$$

* Such transformation is not, in general, unique and can always be found if the known linear relationships are assumed linearly independent, i.e., they are not redundant.

In (6.21), $\dot{y}(t)$ and $z_1(t) = y(t)$ are known from the observation, $z_2(t)$ is to be estimated, and $u_1(t) + u_2(t)$ can be considered as observation noise. Therefore, let us rewrite (6.21) in the more usual form

$$w(t) = -z_2(t) + u_1(t) + u_2(t) \qquad (6.22)$$

where

$$w(t) = \dot{y}(t) + y(t)$$

The processes (6.17) and the observation (6.22) define an estimation problem in terms of a "single process" $z_2(t)$ and an "observed vector" $w(t)$ containing white noise $u_1(t) + u_2(t)$. The solution to this problem is given by (4.50) and (4.51) where the initial condition for the covariance matrix $P(0) = Ez_2{}^2(0) = Ex_2{}^2(0)$. The form of this solution is

$$\begin{aligned}\dot{\hat{z}}_2(t) &= F_1(t)\hat{z}_2(t) + F_2(t)[\dot{y}(t) + y(t)] \\ \hat{z}_2(0) &= 0\end{aligned} \qquad (6.23)$$

where $F_1(t)$ and $F_2(t)$ are obtained in terms of the parameters given by (6.17) and (6.22) and $P(0)$.

The optimal estimator described by (6.23) indicates that we should derive the quantity $\dot{y}(t) + y(t)$ from the observation $y(t)$ and use that as an input to the estimator. This step is not practical to perform, since $\dot{y}(t)$ contains white noise.* A simple change of variable, however, eliminates this problem. Let

$$\hat{h}(t) = \hat{z}_2(t) - F_2(t)y(t) \qquad (6.24)$$

Substituting into (6.23) yields

$$\dot{\hat{h}}(t) = F_1(t)\hat{h}(t) + [F_1(t)F_2(t) + F_2(t) - \dot{F}_2(t)]y(t) \qquad (6.25)$$

In (6.25) only the white-noise-free observation $y(t)$ is used as input to the estimator. However, we note that a term $\dot{F}_2(t)$ has appeared in the equation. This term is easily obtained from (4.47).

Finally it is seen that (6.25) can be used to estimate $h(t)$. Equation (6.24) can then be used to derive $\hat{z}_2(t)$ which along with $\hat{z}_1(t)$ (which is equal to $y(t)$) can be used to determine the desired estimates $\hat{x}_1(t)$ and $\hat{x}_2(t)$ by using (6.15).

▶ **example 2**

Let a scalar signal $s(t)$ be given by

$$s(t) = \alpha e^{-t}$$

* This is not special with this problem and the procedure always leads to this point. However, the cure suggested is always possible.

which is assumed known except for the constant multiplier α. The signal is observed through

$$y(t) = s(t) + n(t)$$

where $n(t)$ is a zero-mean colored noise process which is the output of a double integrator with white noise input. It is desired to estimate $s(t)$ given $y(\tau)$; $0 < \tau \leq t$. The system is modeled by

$$
\begin{aligned}
\dot{x}_1(t) &= x_2(t) \\
\dot{x}_2(t) &= u(t) \\
\dot{x}_3(t) &= -x_3(t) \\
x_3(0) &= \alpha
\end{aligned}
\tag{6.26}
$$

with the observation

$$y(\tau) = x_1(\tau) + x_3(\tau) \qquad 0 \leq \tau \leq t. \tag{6.27}$$

Here, the observation and its first derivative do not contain any white noise as is evident by (6.27) and

$$\dot{y}(t) = \dot{x}_1(t) + \dot{x}_3(t) = x_2(t) - x_3(t) \tag{6.28}$$

while the second derivative is given by

$$\ddot{y}(t) = x_3(t) + u(t) \tag{6.29}$$

The desired change of variable in (6.26) is accomplished by choosing the matrix M in $z = Mx$ as

$$
M = \begin{bmatrix} 1 & 0 & 1 \\ 0 & 1 & -1 \\ 0 & 0 & 1 \end{bmatrix} \tag{6.30}
$$

where the third row is chosen arbitrarily, satisfying only the condition that M be nonsingular. Applying this change of variable leads to the following set of equations

$$\dot{z}_1(t) = z_2(t) \tag{6.31}$$

$$\dot{z}_2(t) = z_3(t) + u(t) \tag{6.32}$$

$$\dot{z}_3(t) = -z_3(t) \tag{6.33}$$

where

$$z_1(t) = x_1(t) + x_3(t) = y(t) \tag{6.34}$$

$$z_2(t) = x_2(t) - x_3(t) = \dot{y}(t) \tag{6.35}$$

Consequently the only quantity to be estimated is $z_3(t)$. From (6.31), (6.32), and (6.34) we can obtain

$$\ddot{y}(t) = z_3(t) + u(t) \tag{6.36}$$

Considering $\ddot{y}(t)$ as the observation, (6.33) and (6.36) can be used to estimate $z_3(t)$ by using (4.50) and (4.51), and $P(0) = Ez_2{}^3(0) = Ex_3{}^2(0) = E\alpha^2$ as initial condition for the variance equation. The estimator is then given by

$$\dot{\hat{z}}_3(t) = F_1(t)\hat{z}_3(t) + F_2(t)\ddot{y}(t) \qquad (6.37)$$

where*

$$F_1(t) = -\frac{3 + \dfrac{2 - P(0)}{P(0)}\,e^{2t}}{1 + \dfrac{2 - P(0)}{P(0)}\,e^{2t}} \qquad (6.38)$$

$$F_2(t) = \frac{2}{1 + \dfrac{2 - P(0)}{P(0)}\,e^{2t}} \qquad (6.39)$$

Again, it is not practical to obtain $\ddot{y}(t)$, since it contains white noise. Hence, let us introduce

$$h(t) = z_3(t) - F_2(t)\dot{y}(t) \qquad (6.40)$$

Substituting into (6.37) yields

$$h(t) = F_1(t)h(t) + [F_1(t)F_2(t) - \dot{F}_2(t)]\dot{y}(t) \qquad (6.41)$$

This is a practical estimator since $\dot{y}(t)$ does not contain white noise and hence can be derived from $y(t)$ by differentiation. The estimate $h(t)$ is then used to obtain $z_3(t)$ from (6.40). Finally $z_3(t)$ along with $z_1(t)$ and $z_2(t)$ which are known exactly (6.34 and 6.35) can be used via the transformation $x = M^{-1}z$ to derive $x_1(t)$, $x_2(t)$, $x_3(t)$. (In this problem $z_3(t) = x_3(t)$ is the ▶ only desired result.)

6.4 Effect of Inaccurate *A Priori* Model Parameters

In order to design a discrete or continuous estimator, we need to have the information specifying the model, namely, the matrices A, B, C, D, K, L, and M and the initial condition for the covariance matrix, namely $P(0)$. In practice, this information may not be very accurate, either due to the lack of exact knowledge concerning the physical processes or as a result of intentionally having used an approximation to the actual processes. Consequently, it is often necessary to determine the effect of the inaccuracies in any of the above parameters on the performance of the estimator. More precisely, we need procedures to help us decide how good an estimator "actually" is.

* For $P(0) < 1$. Similar results can be obtained for $P(0) > 1$.

When the *a priori* information is not correct, the result is that the estimator yields a sub-optimal estimate of $x(k)$ denoted here by $x^*(k)$. Clearly, the covariance matrix of $x(k) - x^*(k)$ (i.e., the actual covariance matrix) is not the same as the covariance matrix which is obtained as a part of the estimator (i.e., the calculated covariance matrix). In other words when equations (4.24) to (4.26) are used with inaccurate values for parameters, the actual covariance matrix of the estimation error may be different from the one calculated by (4.26). In general, one can computationally derive various values of the covariance matrix $P(k)$ for a range of possible *a priori* information (logically in the vicinity of the given data) in order to determine the sensitivity of the covariance function to the data. The continuous estimator can be treated in similar manner. The special case of a discrete estimator with no observation noise $(L = 0)$ and when the incorrect *a priori* data are $P(0)$ is discussed below.

Let the actual covariance matrix of $x(k) - x^*(k)$ be given by

$$P_a(k) = E[x(k) - x^*(k)][x(k) - x^*(k)]' \qquad (6.42)$$

where $x^*(k)$ by definition satisfies the equation

$$x^*(k + 1) = [A - F(k)C]x^*(k) + F(k)y(k) \qquad (6.43)$$

and $F(k)$ is obtained from (4.25) and (4.26). Substituting (4.1) and (6.43) into (6.42) yields

$$P_a(k + 1) = [A - F(k)C]P_a(k)[A - F(k)C]' + BKB' \qquad (6.44)$$

which is identical in form to the calculated covariance matrix equation given by (equation (4.26) with $L = 0$)

$$P(k + 1) = [A - F(k)C]P(k)[A - F(k)C]' + BKB' \qquad (6.45)$$

Defining $E(k) = P(k) - P_a(k)$ and subtracting (6.44) from (6.45) yields

$$E(k + 1) = [A - F(k)C]E(k)[A - F(k)C]'.$$

This equation can be used to determine how the initial error in $P(0)$ propagates through the system. For example, if $E(0) = 0$, that is, if $P(0)$ was correct, then the actual and calculated covariance functions remain the same throughout the process as expected. Furthermore, if $E(0)$ is positive semi-definite, then the right hand side of the equation for $E(k + 1)$ and hence $E(k)$ will be positive semi-definite. Since $E(k)$ is symmetric, this results in all of the elements of the main diagonal of $E(k)$ being non-negative. From the definition of $E(k)$ it means that the actual variances of estimates (which are nonaccessible because accurate information on $P(0)$ is not available) based on incorrect *a priori* information are bounded by the calculated variances (which are accessible) if the initial error matrix $E(0)$ is positive semi-definite.

6.5 Linear Estimators With Constant Memory

In order to mechanize the estimator represented by (4.24) we need to derive values for $F(i)$ and hence $P(i)$ for $i = 0, 1, \ldots, k$. It was pointed out that the equation for the covariance matrix is independent of the observations, and hence values of $P(i)$ can be obtained before any observation is received. In certain applications where it is desired to reduce the amount of computations carried out while the estimator is in operation, equations (4.25) and (4.26) can be solved beforehand and the solution $P(i)$ (or actually $F(i)$) can be stored in the memory of the computer (thus we are not using the recursive method of computing $P(i)$ on-line). We see that such a way of handling $F(i)$ is not in general practical, due to the limitations on the size of memory which can be allocated for this purpose. However, the following approximation to the (Kalman) optimal filter can be used for the case when the parameters specified in (4.1)–(4.5) (A, B, etc.) are time-invariant.

Let us assume that N observation samples are received, namely $y(0), \ldots, y(N - 1)$. Solving (4.24) for $\hat{x}(N)$ yields

$$\hat{x}(N) = \sum_{i=0}^{N-1} \alpha_i y(i) \tag{6.46}$$

where α_i's are constant $n \times m$ matrices obtained from the knowledge of $F(i)$, $i = 0, \ldots, N - 1$. In turn $F(i)$ is obtained from the covariance matrix equation for a given initial condition.

Now suppose the observation $y(N)$ is received. Ordinarily, the optimal estimate $\hat{x}(N + 1)$ would be expressed as a linear function of *all* the past observations $y(i)$, $i = 0, \ldots, N$. However we will, by definition, only use the last N observations to derive $\hat{x}(N + 1)$ using the *same* initial condition for the covariance matrix as the previous step (i.e., let $P(1) = P(0)$). Since this becomes identical with the above problem except that a new set of N observations is used, we have

$$\hat{x}(N + 1) = \sum_{i=0}^{N-1} \alpha_i y(i + 1) \tag{6.47}$$

or finally

$$\hat{x}(k + 1) = \sum_{i=0}^{N-1} \alpha_i y(i + k - N + 1), \qquad k \geq N \tag{6.48}$$

Notice that only the knowledge of matrices α_i, $i = 0, \ldots, N - 1$ are necessary to estimate $\hat{x}(k + 1)$ for $k \geq N$. Clearly the size of memory required is independent of k (i.e., the memory does not grow with k as in the usual case) and need only accommodate the parameters α_i, $i = 0, \ldots, N - 1$ and $y(i + k - N + 1)$, $i = 0, \ldots, N - 1$. As N approaches infinity the estimate given by (6.48) approaches the limit of (4.24). In many practical cases, the simplicity in mechanization of this approximate method justifies its use in place of the optimal estimator.* A measure of the degree of approximation is obtained by comparison of the matrices $P(N)$ and $\lim_{k\to\infty} P(k)$. The comparison simply reveals how much additional error (in the mean square sense) in the estimate $\hat{x}(k)$ is incurred by using the sub-optimal estimator.

▶ **example**

A target is flying on a straight line trajectory with a constant velocity. If $x_1(t)$ is the instantaneous position of the target, we have

$$\dot{x}_1(t) = x_2(t)$$
$$\dot{x}_2(t) = 0$$

The position $x_1(t)$ is observed at 1-second intervals, and the observation contains white noise. The discrete model for the system is then given by

$$x_1(k + 1) = x_1(k) + x_2(k)$$
$$x_2(k + 1) = x_2(k)$$
$$y(k) = x_1(k) + u(k)$$
$$Eu(k_2)u(k_1) = 0.1\,\Delta(k_2 - k_1)$$

We would like to determine the estimate $\hat{x}_1(k + 1)$ from the past three observations [i.e., $y(k), y(k - 1), y(k - 2)$]. Furthermore we are given

$$Ex(0)x'(0) = P(0) = I.$$

Equations (4.25) and (4.26) can be solved for $F(0)$, $F(1)$, and $F(2)$ yielding

$$F(0) = \begin{bmatrix} 0.9 \\ 0 \end{bmatrix}, \qquad F(1) = \begin{bmatrix} 1.75 \\ 0.84 \end{bmatrix}, \qquad F(2) = \begin{bmatrix} 1.27 \\ 0.47 \end{bmatrix}$$

From (4.24) and (6.47) $(\hat{x}(0) = 0)$ the values for α_1, α_2, and α_3 can be obtained:

$$\alpha_1 = \begin{bmatrix} -0.573 \\ -0.343 \end{bmatrix}; \qquad \alpha_2 = \begin{bmatrix} 0.35 \\ 0.014 \end{bmatrix}; \qquad \alpha_3 = \begin{bmatrix} 1.27 \\ 0.47 \end{bmatrix}$$

The estimate of $x(k + 1)$ follows from (6.48)

$$\hat{x}(h + 1) = \alpha_1 y(k - 2) + \alpha_2 y(k - 1) + \alpha_3 y(k)$$

Furthermore, $P(3)$ takes the value

$$P(3) = \begin{bmatrix} 0.2 & 0.09 \\ 0.09 & 0.045 \end{bmatrix}$$

* In fact, the estimator given by (6.48) is the optimal estimator given the data we wish to use.

It is easy to see that, for this particular problem*

$$\lim_{k \to \infty} P(k) = \begin{bmatrix} 0 & 0 \\ 0 & 0 \end{bmatrix}.$$

These results can be interpreted as follows: If the observation has infinite length, then, the variance of, say, $\tilde{x}_1$, is zero. The corresponding variance if the estimator only processes the last three samples is 0.2.

Now if the variance of the estimate values $\hat{x}_1(k)$ and $\hat{x}_2(k)$ as given by $P(3)$ are acceptable, then a very simple constant-coefficient estimator of the ▶ type given by (6.48) has been obtained.

6.6 Numerical Solution of the Error Covariance Equation

From a design point of view, the major step in derivation of the type of linear estimators discussed in Chapter 4 is the solution of the matrix Ricatti equation given by (4.25) and (4.26) in the discrete case and (4.50) in the continuous case. As already pointed out, these are nonlinear equations for which, in general, no practical analytical solutions are available. Resorting to numerical solutions, the matrix Ricatti equation in the discrete form does not introduce any major computational difficulty, since it represents a rather simple recursive relationship.

In the continuous case, equation (4.50) can be simulated on an analog computer. A solution can then be obtained where it will suffer from the inherent inaccuracies associated with solutions on analog computers, especially since we have to use multipliers to simulate the nonlinearity of the equation.

Consequently, it may be more desirable, especially in case of large dimensional systems, to adopt a digital approach. Let us write the matrix equation in question as

$$\dot{P}(t) = f[P(t)] \qquad (6.49)$$

Integrating both sides over the interval $(t_n, t_n + \delta)$ yields

$$P[t_n + \delta] - P(t_n) = \int_{t_n}^{t_n+\delta} f[P(t)] \, dt \qquad (6.50)$$

Since $P(t)$ is the solution of a differential equation and since $f[P(t)]$ is a smooth (quadratic) function of P, the integrand in the right hand side of (6.50) can be approximated by a constant $f[P(t_n)]$ when δ is small. Hence

$$P[t_n + \delta] = P(t_n) + \delta f[P(t_n)] \qquad (6.51)$$

If we denote

$$t_n + \delta = t_{n+1}$$

* In general the limit, if it exists, can be obtained by computational methods.

equation (6.51) results in the following difference equation

$$P[t_{n+1}] = P(t_n) + \delta f[P(t_n)] \tag{6.52}$$

This recursive relationship can be simulated on a digital computer. The value δ is best obtained by trial and error until an acceptable rate of convergence has been achieved. In general, an increase in δ will induce an increase in the rate of convergence up to a certain point. However, it eventually increases the effect of the approximation to a point where the recursive operation may not converge. On the other hand, reducing δ will tend to increase accuracy up to a certain point, beyond which inherent computational accumulated inaccuracies, due to round-off errors on the computer, will become significant.

Equation (6.52) is just one discrete approximation to the differential equation (6.49). Let us define

$$\int f[P(t)] \, dt = F(t)$$

Hence,

$$\int_{t_n}^{t_n+\delta} f[P(t)] \, dt = F(t_n + \delta) - F(t_n)$$

Expanding $F(t_n + \delta)$ in the vicinity of $\delta = 0$ and using the definition of F yields

$$\int_{t_n}^{t_n+\delta} f[P(t)] \, dt = \delta f[P(t_n)] + \frac{\delta^2}{2} \frac{\partial f[P(t_n)]}{\partial P} f[P(t_n)] + \cdots \tag{6.53}$$

It is seen that if we approximate (6.53) by retaining only the first term of the expansion, then (6.51) results. If we keep the first two terms, the discrete approximation to (6.49) becomes

$$P(t_{n+1}) = P(t_n) + \delta f[P(t_n)] + \frac{\delta^2}{2} \frac{\partial f[P(t_n)]}{\partial P} f[P(t_n)] \tag{6.54}$$

Clearly better approximations can be obtained by retaining more terms of the expansion. As we add more terms, the computational aspect of the problem becomes more complex, and the effect of inherent errors such as round-off errors becomes more severe. Consequently, it is rare that an overall improvement is achieved by considering an approximation more complex than (6.54) (i.e., by retaining higher order terms in the expansion).

A serious effect of round-off errors is that, as recursion proceeds, it may tend to produce values for the matrix $P(t_n)$ which represent a nonsymmetric matrix. This in general is not acceptable, and in fact can cause instability of operation (e.g., lack of convergence). The practical approach to avoid this problem is to "symmetrize" the matrix at each step (or every few steps) in the operation by substituting for $P(t_n)$ the matrix $\frac{1}{2}[P(t_n) + P'(t_n)]$.

A simple observation in equation (4.50) reveals that, if in the process the elements of $P(t_n)$ become very small and if K and M are small, the term $f[P(t_n)]$ becomes small, and hence the round-off errors become more critical.

A cure to this problem is in artificially choosing a larger matrix K (for example, by choosing a few or all of its elements larger than the given values). Clearly this will change the problem at hand, and we can only claim a pseudo-optimal estimator. However, this may be acceptable if the alternative is a nonconverging process and hence no estimator at all.

▶ **example**

Let us consider the problem discussed in Section 7 of Chapter 4. The matrix Ricatti equation was a scalar equation given by

$$\dot{P}(t) = -LP^2(t)$$

Let $P(0) = L = 1$, and $\delta = 0.1$. Hence $t_n = 0.1n$, and (6.52) becomes

$$P[0.1(n + 1)] = P[0.1n] + 0.1P^2[0.1n] \qquad (6.55)$$

and equation (6.53) becomes

$$P[0.1(n + 1)] = P[0.1n] - 0.1P^2[0.1n] + 0.01P^3[0.1n] \qquad (6.56)$$

An analytical solution for the covariance equation was also derived for this simple case as

$$P(t) = \frac{1}{t + 1} \qquad (6.57)$$

The following table represents the values of $P(0.1n)$ for $n = 1, \ldots$ corresponding to the exact solution (6.57) and the first and second order approximations given by (6.55) and (6.56).

n	Equation (6.57)	Equation (6.55)	Equation (6.56)
0	1.000	1.000	1.000
1	0.909	0.900	0.910
2	0.834	0.819	0.834
3	0.769	0.752	0.770
4	0.714	0.695	0.716
5	0.666	0.647	0.665

PROBLEMS

1. A scalar random process $y(t)$ has an unknown mean x_1 and a known covariance given by

$$E[y(t_1) - x_1][y(t_2) - x_1] = e^{-|t_2 - t_1|}$$

Derive an estimator for x_1 yielding the estimate $\hat{x}_1$ minimizing $E[x_1 - \hat{x}_1]^2$, given $y(\tau)$, $0 \le \tau \le t$.

2. In Example 1 of Section 6.3, let

$$M = \begin{bmatrix} 1 & 1 \\ 2 & 1 \end{bmatrix}$$

Show that the resulting estimator is identical with one derived in the example.

3. Referring to the process model given by equation (6.1), show that [when $\dot{y}(\tau)$ exists]

$$\min_{\hat{x}(t)} E \, \|x(t) - \hat{x}(t)\|^2 \,|\, _{y(\tau),0 \leq \tau \leq t} = \min_{\hat{x}} E \, \|x(t) - \hat{x}(t)\|^2 \,|\, _{y(\tau),\dot{y}(\tau),0 \leq \tau \leq t}$$

4. In the problem of estimating the mean of a white noise sequence (Section 4.4), let the true value of *a priori* variance of the mean and the value used by the estimator be 1 and 2, respectively. Discuss the effect of this inaccuracy in "given data" on the performance of the estimator.

5. Let $y(k)$ be a random sequence with mean $x_1 = x_1(0)e^{-k}$, $k > 0$ and

$$E[y(k_1) - x_1][y(k_2) - x_1] = \Delta(k_2 - k_1)$$

Design a "constant memory" estimator for estimating the parameter $x_1(0)$ resulting in estimation error variance less than 1. The observation consists of the values $y(k)$, $k \geq 0$, and the *a priori* variance of $x_1(0)$ is given by

$$Ex_1^2(0) = 10.$$

SELECTED READINGS

Bryson, A. E. and D. E. Johansen, "Linear Filtering for Time-Varying Systems Using Measurements Containing Colored Noise," *IEEE Transactions on Automatic Control*, January 1965.

Kalman, R. E., "A New Approach to Linear Filtering and Prediction Problems," *J. Basic Engineering, ASME Transactions*, vol. 82D March 1960.

Kalman, R. E. and R. S. Bucy, "New Results in Linear Filtering and Prediction Theory," *J. Basic Engineering, ASME Transactions*, March 1961.

Mehra, R. K. and A. E. Bryson, *Smoothing for Time-Varying Systems Using Measurements Containing Colored Noise*, Division of Engineering and Applied Physics, Harvard University, Cambridge, Massachusetts, June 1967.

Nishimura, T., "On the a Priori Information in Sequential Estimation Problems," *IEEE Transactions on Automatic Control*, Vol. AC-11, No. 2, April 1966.

Soong, T. T., "On a Priori Statistics in Minimum Variance Estimation Problems," *Trans. ASME, Ser. D., J. Basic Engineering*, **87**, 1965.

chapter 7

NONLINEAR ESTIMATION

7.1 Introduction

The problem of estimating the parameters or states of a nonlinear system—whether the nonlinearity is introduced by the model generating the stochastic processes or by the observation mechanism—is a very complicated one and by no means is solved in usable form in the general case. The practical need for solutions to such problems has resulted in a large number of ideas and procedures. Some are no more than a philosophy of approach rather than a procedure leading to derivation of practical estimators. There are a few procedures which attack a specific problem and result in useful estimators satisfying the limited objectives. In general, one has to accept that an analytical solution in closed form is not likely to be available, and computational algorithms should be sought in their place. In fact, one may arrive at this conclusion even in the case of general linear estimators discussed in Chapters 4 and 5.

Due to the success in linear estimation, it is logical to attempt the application of related procedures to a class of nonlinear systems whose behavior is close to that of linear systems. Clearly, one can at best expect to derive an estimator which is only approximately optimal. The use of linear estimation procedures for nonlinear problems is the first topic introduced in this chapter. The remainder of this chapter is devoted to the characterization of the general nonlinear estimation process from a Bayesian point of view (i.e., minimization of an average cost requiring *a priori* knowledge of the probability density function on the parameter to be estimated). The procedures of maximum likelihood estimation and least-square estimation which usually require no *a priori* knowledge of the parameter probability density function will be discussed in the next chapter.

Since we are searching for numerical algorithms for the solution to the estimation problem, discrete processes will mainly be considered in this chapter. Consequently, if we have to deal with continuous systems, a suitable discrete approximation should be employed (*see* Chapter 3).

7.2 Problem Statement

Let the vector-valued random sequence $x(k)$ be given by the following (nonlinear) difference equation model.

$$x(k + 1) = f[x(k), k] + Bu(k) \tag{7.1}$$

$$Eu(k) = 0$$

$$Eu(k_1)u'(k_2) = K \, \Delta[k_2 - k_1] \tag{7.2}$$

In (7.2), K is the $r \times r$ covariance matrix, which is in general a function of k. The Δ-notation in (7.2) implies that the noise sequence is white, i.e., successive noise terms are uncorrelated. Let the observations be given by

$$y(k) = g[x(k), k] + Dv(k) \tag{7.3}$$

$$Ev(k) = 0$$

$$Ev(k_1)v'(k_2) = L \, \Delta(k_2 - k_1) \tag{7.4}$$

where L is an $q \times q$ covariance matrix which is also, in general, a function of k. The random sequences $u(k)$ and $v(k)$ in this model are uncorrelated sequences* which in general need not be gaussian. Notice that the random noise terms due to $u(k)$ and $v(k)$ have been chosen to appear in additive form. (A more general model would include the noise as arguments of the nonlinear functions f and g.) It is desired to estimate $x(k + 1)$ by processing the observed data $y(i)$, $0 \leq i \leq k$, and satisfying some specified criterion of estimation.

7.3 Method of Linearization

The philosophy here is to determine relationships representing the behavior of solutions $x(k)$ of equations (7.1) and (7.3) in the vicinity of a nominal

---

* If this "uncorrelatedness" condition is not satisfied for a practical model, usually the correlated random inputs to the difference equation can be represented as outputs of another system excited by uncorrelated (white) noise. The model given by (7.1) and (7.2) can then represent the total (or "augmented") system.

solution denoted by $x^*(k)$. In many practical cases this behavior (i.e., the dynamics of the difference $x(k) - x^*(k)$) can be characterized by a set of linear equations. Consequently, the ordinary Kalman filtering of Chapter 4 can be used to compute an estimate $\hat{x}(k)$ of $x(k)$ in a neighborhood of $x^*(k)$. Now if $x^*(k)$ is close to the true value of $x(k)$ the above procedure should give us a reasonably good estimate. Consequently, the choice of the nominal solution $x^*(k)$ is very important. In some applications, an approximate knowledge of the true $x(k)$ is available, which is then chosen as the candidate for $x^*(k)$. Such is the case in orbit estimation of satellites where, usually, an approximate orbit is known *a priori*. On the other hand, the "Kalman" estimate $\hat{x}(k)$ is also hopefully a "good" estimate of the true $x(k)$ and can logically be used as a candidate for the "nominal" solution $x^*(k)$.

Let us assume the functions $f(x(k), k)$ and $g(x(k), k)$ are twice continuously differentiable in $x(k)$. Furthermore, let us define two matrices $A(k)$ and $C(k)$ which are $n \times n$ and $q \times n$, respectively, and are functions of k:

$$A(k) \triangleq \begin{bmatrix} \dfrac{\partial f_1}{\partial x_1} & \dfrac{\partial f_1}{\partial x_2} & \\ \dfrac{\partial f_2}{\partial x_1} & \ddots & \\ & & \dfrac{\partial f_n}{\partial x_n} \end{bmatrix}_{x(k) = x^*(k)} \tag{7.5}$$

$$C(k) \triangleq \begin{bmatrix} \dfrac{\partial g_1}{\partial x_1} & \dfrac{\partial g_1}{\partial x_2} & \\ \dfrac{\partial g_2}{\partial x_1} & \ddots & \\ & & \dfrac{\partial g_q}{\partial x_n} \end{bmatrix}_{x(k) = x^*(k)} \tag{7.6}$$

In (7.5, 6), f_i and g_i are the ith components of the vectors $f(x(k), k)$ and $g(x(k), k)$, respectively. We shall use these A and C matrices as the coefficient matrices of a *linearized* representation of (7.1) and (7.3) in the neighborhood of the nominal solution $x^*(k)$.

The deviation (sometimes called the "perturbation") of the nominal solution $x^*(k)$ from the true value $x(k)$ will be denoted by $x^e(k)$. Hence, in symbols

$$x^e(k) \triangleq x(k) - x^*(k) \tag{7.7}$$

Naturally, we also have $x^e(k + 1) \triangleq x(k + 1) - x^*(k + 1)$.

Expanding the right hand sides of (7.1) and (7.3) in the vicinity of $x^*(k)$ and dropping the nonlinear terms (i.e., terms of higher order than the first) in $x^e(k)$ yields

$$x(k + 1) = f[x^*(k), k] + A(k)x^e(k) + Bu(k) \qquad (7.8)$$

$$y(k) = g[x^*(k), k] + C(k)x^e(k) + Dv(k) \qquad (7.9)$$

In (7.8, 9) the vector functions f and g at each k are now assumed known, since the known "nominal" solution has been substituted. Also, the matrices A and C obtained by differentiation are now assumed known, since the derivatives are evaluated at the nominal solution $x^*(k)$. Substituting for $x(k + 1)$ from (7.7) into (7.8) we obtain

$$x^e(k + 1) = A(k)x^e(k) + Bu(k) + f[x^*(k), k] - x^*(k + 1) \qquad (7.10)$$

Let us define

$$h(k) \triangleq y(k) - g(x^*(k), k) \qquad (7.11)$$

Substituting (7.11) into (7.9) yields

$$h(k) = C(k)x^e(k) + Dv(k) \qquad (7.12)$$

Since $x^*(k)$ is assumed known, the quantity $h(k)$ can be obtained from the knowledge of observation $y(k)$. Therefore, except for the known input terms $f[x^*(k), k] - x^*(k + 1)$ to the difference equation described by (7.10), the linear model given by (7.10) and observations given by (7.12) are of the same form as equations (4.1) and (4.2) of Chapter 4.* The effects of known inputs are easily taken into account by eliminating their contributions to $x^e(k)$ and $h(k)$. This is possible because (7.10) and (7.12) are linear systems and consequently the principle of superposition is applicable.

Let $x^f(k)$ be defined as the solution to the linear difference equation (7.10) with zero initial conditions and when $Bu(k)$ is set equal to zero. That is

$$x^f(k + 1) = A(k)x^f(k) + f[x^*(k), k] - x^*(k + 1) \qquad (7.13)$$

We introduce the variable $z(k)$ which is defined to be equal to x^e in (7.10) minus the contribution of the known input:

$$z(k) = x^e(k) - x^f(k) \qquad (7.14)$$

Subtracting (7.13) from (7.10) and making use of (7.14) results in

$$z(k + 1) = A(k)z(k) + Bu(k) \qquad (7.15)$$

And from (7.11), (7.12), and (7.14) it follows that

$$y(k) - g(x^*(k), k) - C(k)x^f(k) = C(k)z(k) + Dv(k) \qquad (7.16)$$

---

* As will be seen, only the knowledge of $x * (i)$, $0 \le i \le k$, is necessary to derive the estimate $\hat{x}(k + 1)$. This is essential if we want to choose $x * (k) = \hat{x}(k)$.

The left hand side of (7.16) is known and can be appropriately called the modified observation. Equations (7.15) and (7.16) are then identical with (4.1) and (4.2) of Chapter 4. Consequently, the problem of deriving the estimate $z(k + 1)$ (denoted by $\hat{z}(k + 1)$) given $y(i)$ and $x^*(i)$, $0 \le i \le k$, has been solved. Notice that for this estimate we only need values of $x^f(i)$ for $0 \le i \le k$ which from (7.13) only requires knowledge of $x^*(i)$, $0 \le i \le k$. Finally from (7.7) and (7.14) it follows that

$$\hat{x}(k + 1) = \hat{z}(k + 1) + x^f(k + 1) + x^*(k + 1) \tag{7.17}$$

Equation (7.17) yields the desired estimate $\hat{x}(k + 1)$ if $x^*(k + 1)$ is known. If we decide to choose $x^*(k) = \hat{x}(k)$, then clearly (7.17) is not a meaningful relationship. In this important special case, it follows from (7.7) that

$$\hat{x}^e(k) = \hat{x}(k) - x^*(k) = 0$$

and consequently, from (7.13) and (7.14), we have the simple and very useful relationship

$$\hat{x}(k + 1) = f[\hat{x}(k), k] + \hat{z}(k + 1) - A(k)\hat{z}(k) \tag{7.18}$$

A practical advantage of choosing the nominal solution $x^*(k)$ to be equal to $\hat{x}(k)$ lies in the fact that it is no longer required to store all the values of $x^*(k)$ in the estimator computer memory. This is because, at each k, the estimate of $\hat{x}(k + 1)$ depends only on the last estimate $\hat{x}(k)$, rather than on some stored values $x^*(k)$, $x^*(k - 1)$, etc. The procedure is as follows: Starting with $\hat{x}(k)$, $\hat{z}(k)$, and $x^f(k)$, the matrices $A(k)$ and $C(k)$ are evaluated from (7.5) and (7.6) with $x^*(k) \triangleq \hat{x}(k)$. Next $\hat{z}(k + 1)$ is obtained by solving the estimation problem associated with (7.15) and (7.16). Then (7.18) is used to derive $\hat{x}(k + 1)$ which is the desired result. In addition, (7.13) will be used to obtain $x^f(k + 1)$ needed for the succeeding step. We may proceed further and apply the results expressed by (4.24) to estimate $z(k + 1)$ in (7.15) and (7.16). It follows that

$$\hat{z}(k + 1) - A(k)\hat{z}(k) = F(k)\{y(k) - g[\hat{x}(k), k]\}$$

Hence, substituting into (7.18) yields

$$\hat{x}(k + 1) = f[\hat{x}(k), k] + F(k)\{y(k) - g[\hat{x}(k), k]\} \tag{7.19}$$

where $F(k)$ satisfies (4.25) and (4.26) with matrices A and C defined by (7.5) and (7.6).

The continuous counterpart of the discrete estimator in (7.19) can be derived in similar manner. The results are given below and the derivation is left as an exercise.

Let the vector-valued random process $x(t)$ and observation vector $y(t)$ be given by the following nonlinear differential equation model.

$$\dot{x}(t) = f[x(t), t] + Bu(t)$$
$$y(t) = g[x(t), t] + Dv(t)$$

where

$$Eu(t) = Ev(t) = 0$$
$$Eu(t_1)u'(t_2) = K\,\delta(t_2 - t_1)$$
$$Ev(t_1)v'(t_2) = L\,\delta(t_2 - t_1)$$
$$Eu(t_1)v'(t_2) = M\,\delta(t_2 - t_1)$$

The estimate of $x(t)$ given $y(\tau)$, $0 \le \tau \le t$ is denoted by $\hat{x}(t)$ and satisfies the following nonlinear differential equation,

$$\dot{\hat{x}}(t) = f[\hat{x}(t), t] + [P(t)C' + \text{BMD}'][DLD']^{-1}\{y(t) - g[\hat{x}(t), t]\}$$
$$\hat{x}(0) = 0$$

where $P(t)$ satisfies (4.50) with matrices A and C defined by

$$A = \left.\frac{\partial f[x(t), t]}{\partial x(t)}\right|_{x=\hat{x}(t)}$$

$$C = \left.\frac{\partial g[x(t), t]}{\partial x(t)}\right|_{x=\hat{x}(t)}$$

▶ **example**

Let the scalar random process $x(k)$ be defined by

$$x(k + 1) = -0.5x(k) - 0.1x^3(k) + u(k)$$
$$Eu(k) = 0$$
$$Eu(k_1)u(k_2) = 0.1\,\Delta(k_2 - k_1)$$
$$Ex(0) = 0$$
$$Ex^2(0) = 1$$

It is desired to determine the estimate $\hat{x}(k + 1)$ minimizing $E[x(k + 1) - \hat{x}(k + 1)]^2$ by processing the observed data $y(i)$, $0 \le i \le k$, where

$$y(k) = x(k) + v(k)$$
$$Ev(k) = 0$$
$$Ev(k_1)v(k_2) = \Delta(k_2 - k_1)$$

Choosing to use the estimate $\hat{x}(k)$ for the nominal solution $x * (k)$, we may evaluate the matrices $A(k)$ and $C(k)$ from (7.5, 6) to obtain

$$A(k) = -0.5 - 0.3\hat{x}^2(k)$$
$$C(k) = 1$$

We may now directly proceed to utilize (7.19) and evaluate the only remaining term, $F(k)$, from (4.25) and (4.26). However, in the following the intermediate steps are shown for illustration.

Equation (7.13) becomes

$$x^f(k+1) = [-0.5 - 0.3\hat{x}^2(k)]x^f(k) - 0.5\hat{x}(k) - 0.1\hat{x}^3(k) - \hat{x}(k+1)$$

and (7.15) and (7.16) yield

$$z(k+1) = [-0.5 - 0.3\hat{x}^2(k)]z(k) + u(k)$$
$$y(k) - \hat{x}(k) - x^f(k) = z(k) + v(k)$$

From the results of Chapter 4, the optimal estimator for $z(k)$ is then (equations 4.24–4.26)

$$\hat{z}(k+1) = [-0.5 - 0.3\hat{x}^2(k) - F(k)]\hat{z}(k) + F(k)[y(k) - \hat{x}(k) - x^f(k)]$$

$$F(k) = \frac{-0.5 - 0.3\hat{x}^2(k)}{1 + P(k)} P(k)$$

$$P(k+1) = [-0.5 - 0.3\hat{x}^2(k) - F(k)]^2 P(k) + 0.1 + F^2(k)$$

with the following initial conditions

$$x^f(0) = \hat{z}(0) = \hat{x}(0) = 0; \qquad P(0) = 1$$

Finally, (7.18) yields the desired estimate given by

$$\hat{x}(k+1) = \hat{z}(k+1) + 0.5\hat{z}(k) - 0.5\hat{x}(k) + 0.3\hat{x}^2(k)\hat{z}(k) - 0.1\hat{x}^3(h)$$

or equivalently by (utilizing (7.19))

▶ $$\hat{x}(k+1) = -0.5\hat{x}(k) - 0.1\hat{x}^3(k) + F(k)[y(k) - \hat{x}(k)]$$

We note in conclusion that we have relied heavily on linearization of the f and g functions to obtain the matrices $A(k)$ and $C(k)$. Further, we have relied on the fact that the noise terms $u(k)$ and $v(k)$ contributed additively in equation (7.13). If the functions f and g had been "less smooth, "the values of the matrices $A(k)$ and $C(k)$ might have fluctuated severely when the nominal trajectory $x^*(k)$ was substituted instead of the exact trajectory $x(k)$. It is also to be noted that the equation for $P(k)$, and hence $F(k)$, depends on $\hat{x}(k)$ and, consequently, it can not be solved in advance of receiving the observations.

7.4 Evolution of A Posteriori Probability Density Function

In the general nonlinear estimation problem characterized from a Bayesian point of view (i.e., when we would like to minimize the average (or "expected") value of a given cost function), a central role is played by the a posteriori probability density function. This is the probability density function on the parameter to be estimated, conditioned on the received observations which, for the estimation problem considered in the previous section, may be symbolized as

$$p[x(k + 1) \mid y(0), \ldots, y(k)] \qquad (7.20)$$

If this density function were available, we could obtain a number of estimates for $x(k + 1)$ satisfying different criteria of optimality such as minimum quadratic cost, conditional expectation [which is the mean of (7.20)], the mode of (7.20),* etc. (*See* Chapter 2). These various estimates are "functionals" of the entire density function (7.20), and could be obtained by straightforward computational manipulation of the density function.

Now, the major problem in estimating $x(k + 1)$ is the derivation of the above posterior density function (7.20). Since in general this probability density function is not gaussian, it is not sufficient, say, to determine the density by evaluating its first two moments (or in general any finite number of moments†). We shall seek feasible methods for determining (7.20) in various cases. In particular, we shall consider a method whereby we could at least evaluate (7.20) (for each k) in a sequential (recursive) manner.

It happens for the problem considered in the previous section, that the following probability density functions can be directly evaluated from the knowledge of the probability density functions on $u(k)$ and $v(k)$.

$$p[x(k + 1) \mid x(k)] \qquad (7.21)$$

$$p[y(k) \mid x(k)] \qquad (7.22)$$

One can see, from (7.1, 3), that the density function defined by (7.21) is that of the random variable $Bu(k)$ with the mean shifted by $f(x(k), k)$, and that the function defined by (7.22) is the density function of $Dv(k)$ with the mean shifted by $g(x(k), k)$. Clearly, these density functions are completely defined by their first and second moments, given by (7.2) and (7.4), when $u(k)$ and $v(k)$ are gaussian.

The joint probability density function $p[x(k + 1), y(k) \mid y(0), \ldots, y(k - 1)]$ can be simply written as the identity

$$p[x(k + 1), y(k) \mid y(0), \ldots, y(k - 1)]$$
$$\triangleq \int p[x(k + 1), x(k), y(k) \mid y(0), \ldots, y(k - 1)]\, dx(k) \qquad (7.23)$$

---

* The *a posteriori* mode is not in fact a Bayes estimate. However, it is defined by the conditional density given by (7.20).
† In general, a derivation of any of the various moments of some desired probability density function is quite difficult when we do not know the form of the density function in advance.

The integrand of the right hand side of (7.23) can be expressed* in terms of (7.21), (7.22), and $p[x(k) \mid y(0), \ldots, y(k-1)]$:

$$p[x(k+1), x(k), y(k) \mid y(0), \ldots, y(k-1)]$$
$$= p[y(k) \mid x(k+1), x(k), y(0), \ldots, y(k-1)]$$
$$\times p[x(k+1), x(k) \mid y(0), \ldots, y(k-1)] \quad (7.24)$$

Since $v(k)$ is an independent sequence, we have

$$p[y(k) \mid x(k+1), x(k), y(0), \ldots, y(k-1)] = p[y(k) \mid x(k)] \quad (7.25)$$

The last term in (7.24) can be expressed equivalently by

$$p[x(k+1), x(k) \mid y(0), \ldots, y(k-1)]$$
$$= p[x(k+1) \mid x(k), y(0), \ldots, y(k-1)]p[x(k) \mid y(0), \ldots, y(k-1)]$$
$$(7.26)$$

Now, since $u(k)$ also is an independent sequence, we have

$$p[x(k+1) \mid x(k), y(0), \ldots, y(k-1)] = p[x(k+1) \mid x(k)] \quad (7.27)$$

Using Bayes' formula (Section 1.24) for manipulation of probability densities to obtain the posterior probability, we get

$$p[x(k+1) \mid y(0), \ldots, y(k)] = \frac{p[x(k+1), y(k) \mid y(0), \ldots, y(k-1)]}{p[y(k) \mid y(0), \ldots, y(k-1)]}$$
$$(7.28)$$

Substituting for the terms in the right hand side of (7.28) from (7.23)–(7.27) yields

$$p[x(k+1) \mid y(0), \ldots, y(k)]$$
$$= \frac{\int p[y(k) \mid x(k)]p[x(k+1) \mid x(k)]p[x(k) \mid y(0), \ldots, y(k-1)]\,dx(k)}{\int p[y(k) \mid x(k)]p[x(k) \mid y(0), \ldots, y(k-1)]\,dx(k)}$$
$$(7.29)$$

Equation (7.29) is the desired result. Starting with a probability density on the initial condition, $p[x(0)]$, (which should be provided *a priori*) and the knowledge of the conditional densities given by (7.21) and (7.22), the *a posteriori* probability density $p[x(k+1) \mid y(0), \ldots, y(k)]$ may be obtained from $p[x(k) \mid y(0), \ldots, y(k-1)]$ and the observation $y(k)$.

---

* See Bayes' formula, equation (1.120).

In the above derivation of (7.29) it was assumed that the random terms $Bu(k)$ and $Dv(k)$ in (7.1) and (7.3) were additive as indicated. This assumption can be easily eliminated from the derivations, thereby leading to results very similar to (7.29). Although (7.29) represents a simple recursive relationship, the integrations in the right hand side in general can only be carried out computationally, and they represent routine but tedious operations except in very special cases such as scalar systems. When $u(k)$, $v(k)$, and $p[x(0)]$ are assumed to be gaussian, (7.29) will be simplified somewhat. In fact, if we choose as the estimate of $x(k)$ the conditional expectation

$$\hat{x}(k) = E[x(k) \,|\, y(0), \ldots, y(k-1)] \tag{7.30}$$

and furthermore define the covariance matrix $P(k)$ by

$$P(k) = E[x(k) - \hat{x}(k)][x(k) - \hat{x}(k)]' \tag{7.31}$$

and if we use the linear model given by equations (4.1)–(4.5) in place of (7.1)–(7.4), then equation (7.29) can be used to derive the Kalman filter obtained in Chapter 4. Since the purpose here is not to re-derive the Kalman filter, this procedure is outlined below by means of a very simple example. The main intention here is to familiarize the reader with the computational steps involved in using (7.29) in general form (however, in this example, the computational steps are simple enough to be carried out analytically).

▶ **example**

It is desired to estimate the mean (denoted by α) of a scalar white gaussian process $y(k)$ from the observations $y(0), \ldots, y(k)$. The model of the system is given by

$$x(k+1) = x(k); \quad x(0) = \alpha \tag{7.32}$$

$$y(k) = x(k) + v(k); \quad Ev(k) = 0 \tag{7.33}$$

$$Ev(k_1)v(k_2) = L\Delta(k_2 - k_1) \tag{7.34}$$

Since there is no random term in (7.32), it follows that $p[x(k+1) \,|\, x(k)] = \delta[x(k+1) - x(k)]$. Consequently, equation (7.29) can be simply written as follows*

$$
p[x(k+1) \,|\, y(0), \ldots, y(k)]
$$
$$
= \frac{\{p[y(k) \,|\, x(k)]p[x(k) \,|\, y(0), \ldots, y(k-1)]\} \,|\, x(k) = x(k+1)}{p[y(k) \,|\, y(0), \ldots, y(k-1)]}
$$
$$\tag{7.35}$$

Clearly, all of the probability functions in (7.35) are gaussian.

---

*In the computational version, $\int p[y(k) \,|\, x(k)]p[x(k) \,|\, y(0), \ldots, y(k-1)] \, dx(k)$ should be used in place of its equivalent value $p[y(k) \,|\, y(0), \ldots, y(k-1)]$ as indicated by (7.29). The change is made here to avoid a routine but complicated integration.

From (7.30) and (7.33) we have

$$E[y(k) \mid y(0), \ldots, y(k-1)] = \hat{x}(k) \qquad (7.36)$$

and from (7.31)

$$E[y(k) - \hat{x}(k)]^2 = P(k) + L \qquad (7.37)$$

Also, from (7.33)

$$E[y(k) \mid x(k)] = x(k) \qquad (7.38)$$

$$E[y(k) - x(k)]^2 = L \qquad (7.39)$$

Clearly, $x(k+1)$ is a gaussian vector. By definition (equation 7.30), it has mean $\hat{x}(k+1)$, and from (7.31) its covariance is $P(k+1)$. Consequently, (7.35) can be written analytically as follows

$$C_1 \exp -\frac{1}{2} \frac{[x(k+1) - \hat{x}(k+1)]^2}{P(k+1)}$$

$$= C_2 \exp -\frac{1}{2} \left\{ \frac{[y(k) - x(k+1)]^2}{L} + \frac{[x(k+1) - \hat{x}(k)]^2}{P(k)} \right.$$

$$\left. - \frac{[y(k) - \hat{x}(k)]^2}{P(k) + L} \right\} \qquad (7.40)$$

By rearranging terms, (7.40) can easily be rewritten as follows

$$C_1 \exp -\frac{1}{2} \frac{[x(k+1) - \hat{x}(k+1)]^2}{P(k+1)}$$

$$= C_2 \exp -\frac{1}{2} \frac{\left[x(k+1) - \dfrac{L}{P(k) + L} \hat{x}(k) - \dfrac{P(k)}{P(k) + L} y(k) \right]^2}{\dfrac{LP(k)}{L + P(k)}} \qquad (7.41)$$

where C_1 and C_3 are constants independent of $x(k+1)$. Equation (7.41) is by definition an identity in terms of $x(k+1)$. Hence, equating the mean and variance of both sides yields

$$P(k+1) = \frac{LP(k)}{L + P(k)} \qquad (7.42)$$

$$\hat{x}(k+1) = \frac{L}{P(k) + L} \hat{x}(k) + \frac{P(k)}{P(k) + L} y(k) \qquad (7.43)$$

The results given by (7.42) and (7.43) are identical with the Kalman filter derived for the same estimation problem in the example of Section 4.4. The desired initial condition $P(0)$ in (7.42) is simply the variance of ▶ $p(x(0))$.

7.5 Nonlinear Recursive Estimation, The Direct Method

In the preceding section a computational procedure was introduced for obtaining the *a posteriori* probability density function of the parameter to be estimated. The procedure was of a recursive nature and consequently, regardless of the form of estimation criterion chosen (as long as it can be evaluated from the *a posteriori* density function), we have a recursive estimator. The procedure, however, has practical shortcomings. Let us assume we want to choose the mode of the *a posteriori* density function as our estimate. At each step we have to (a) find the density *function* $p[x(k+1) \mid y(0), \ldots, y(k)]$, (b) search for its peak, and (c) keep the entire function in the computer memory to be used in the succeeding step. From a practical point of view, it would be more satisfactory if we could derive a recursive relationship operating on the estimate (in this case, the posterior mode) itself rather than the *a posteriori* density function. In general, this represents a very complicated problem.* In the special case when the density function is gaussian, since the first two moments describe the density function completely, we can expect certain simplifications. However, as will be obvious shortly, we simply substitute one tedious computational procedure for another.

Let the vector-valued random process $x(k)$ and observation $y(k)$ be given by (7.1)–(7.4) repeated below†

$$x(k+1) = f[x(k)] + Bu(k) \qquad (7.44)$$

$$y(k) = g[x(k)] + Dv(k) \qquad (7.45)$$

with $u(k)$ and $v(k)$ being independent gaussian sequences having the covariance functions K and L, respectively.

It is desired to estimate the entire sequence $x(0), \ldots, x(k)$ by obtaining the mode of the *a posteriori* probability density

$$p[x(0), \ldots, x(k) \mid y(0), \ldots, y(k)] \qquad (7.46)$$

This is slightly different from the problem we have considered previously, where we usually obtained the estimate for $x(k)$ or $x(k+1)$ given the observed data $y(0), \ldots, y(k)$. Here we search for the estimate of the entire trajectory (the sequence $x(0), \ldots, x(k)$). Only in the case of linear systems

---

* In the case of continuous processes, and under certain conditions, an exact *stochastic* differential equation for the mode of the *a posteriori* density has been obtained by Kushner. This represents an interesting result. However, thus far it offers no solution to practical problems owing to monumental computational difficulties.

† For notational simplicity, the terms f, g, B, D, K, and L, do not give the dependence on k explicitly. The procedure, however, is the same in the general case.

can one show that the optimal (posterior mode) estimate of $x(k)$ (and consequently $x(i)$, $i > k$) remains the same whether we determine the mode of $p[x(k) \mid y(0), \dots, y(k)]$ or one given by (7.46). This modification of the problem is necessary for the following derivations.

The function represented by the conditional density in (7.46) is extremely complex.* Using Bayes' rule it can be represented by

$$p[x(0), \dots, x(k) \mid y(0), \dots, y(k)]$$

$$= \frac{p[y(0), \dots, y(k) \mid x(0), \dots, x(k)]p[x(0), \dots, x(k)]}{p(y(0), \dots, y(k))} \quad (7.47)$$

After the observations $y(0), \dots, y(k)$ have been received, the denominator in the right hand side is a number independent of $x(0), \dots, x(k)$. Consequently, maximization (to find the mode) of the left hand side of (7.47) with respect to $x(0), \dots, x(k)$ is equivalent to the following operation:

$$\max_{x(0), \dots, x(k)} p[y(0), \dots, y(k) \mid x(0), \dots, x(k)]p[x(0), \dots, x(k)] \quad (7.48)$$

The terms in (7.48) can be readily obtained from the system description. Since the sequence $v(k)$ is an independent sequence, from (7.45), we have

$$p[y(0), \dots, y(k) \mid x(0), \dots, x(k)] = \prod_{i=0}^{k} p_L[y(i) - g(x(i))] \quad (7.49)$$

where p_L is a gaussian density function with zero mean and covariance function given by DLD'.

The joint density function $p[x(0), \dots, x(k)]$ can be written as

$$p[x(0), \dots, x(k)] = p[x(0)] \prod_{i=1}^{k} p[x(i) \mid x(i-1)] \quad (7.50)$$

Using (7.44), equation (7.50) yields

$$p[x(0), \dots, x(k)] = p[x(0)] \prod_{i=1}^{k} p_K[x(i) - f(x(i-1))] \quad (7.51)$$

where p_K is a gaussian density function with zero-mean and covariance matrix given by BKB'. The density $p[x(0)]$ is a gaussian density with zero mean and covariance $P(0)$ which is to be given as a part of the input data.

* Except, of course, for the linear process, where this probability density function becomes gaussian.

Using (7.49) and (7.51), and for the moment assuming BKB' to be non-singular, the operation in (7.48) can be written as

$$\max_{x(0),\,\ldots,\,x(k)} C \exp -\frac{1}{2}\Bigg\{ \sum_{i=0}^{k} [y(i) - g[x(i)]]'[DLD]^{-1}[y(i) - g[x(i)]]$$

$$+ x'(0)P^{-1}(0)x(0) + \sum_{i=1}^{k} [x(i) - f[x(i-1)]]'[BKB']^{-1}$$

$$\times [x(i) - f[x(i-1)]]\Bigg\}$$

where C is a constant independent of $x(0), \ldots, x(k)$. Clearly, this maximization leads to minimization of the exponent. Using (7.44), this leads to

$$\min_{\substack{x(0),\,\ldots,\,x(k) \\ u(0),\,\ldots,\,u(k-1)}} \sum_{i=0}^{k} [y(i) - g[x(i)]]'[DLD']^{-1}[y(i) - g[x(i)]] + x'(0)P^{-1}(0)x(0)$$

$$+ \sum_{i=1}^{k} u'(i-1)K^{-1}u(i-1) \quad (7.52)$$

where the minimization is subject to the constraint

$$x(k+1) = f[x(k)] + Bu(k) \quad (7.53)$$

The minimization suggested by (7.52) may be used when BKB' is singular as long as K is positive definite.

It is clear that the original estimation problem is replaced by a deterministic problem, namely a functional minimization (equation (7.52)) subject to a constraint specified by the difference equation (7.53). A number of techniques have been suggested for solving this problem, none of which offers a simple or even practical procedure. One such approach which is somewhat reasonable is discussed in the following.

Let us define the optimal estimate $\hat{x}(k)$ as the last term in the sequence $x(0), \ldots, x(k)$, where this sequence satisfies (7.52) and (7.53). A further assumption is made that (7.44) can be solved for $x(k)$, i.e., we can find a function denoted by f^{-1} such that we can indicate the solution for $x(k)$ by

$$x(k) = f^{-1}[x(k+1) - Bu(k)] \quad (7.54)$$

A special function $S_k[x(k)]$ is defined as follows

$$S_k[x(k)] = \min_{u(0),\,\ldots,\,u(k-1)} \Bigg\{ \sum_{i=0}^{k} [y(i) - g[x(i)]]'[DLD']^{-1}[y(i) - g(x(i))]$$

$$+ x'(0)P^{-1}(0)x(0) + \sum_{i=1}^{k} u'(i-1)K^{-1}u(i-1) \Bigg\} \quad (7.55)$$

where $x(i)$, $0 \leq i \leq k-1$, are substituted by their values in terms of $x(k)$ and $u(0), \ldots, u(k-1)$ by utilizing (7.45). Consequently, $S_k[x(k)]$ is a function of $x(k)$ only. Using (7.55) the estimation of $x(k)$ leads to

$$\min_{x(k)} S_k[x(k)] \qquad (7.56)$$

Separating the minimization with respect to $u(k-1)$ from the remaining terms in (7.55) yields

$$S_k[x(k)] = \min_{u(k-1)} \{[S_{k-1}[x(k-1)]] + [y(k) - g(x(k))]'[DLD']^{-1}$$
$$\times [y(k) - g(x(k))] + u'(k-1)K^{-1}u(k-1)\} \qquad (7.57)$$

Substituting for $x(k-1)$ from (7.54) into (7.57), we may derive a functional equation yielding the optimal estimate $\hat{x}(k)$.

$$\min_{x(k)} S_k(x(k)) = \min_{x(k)} \Big\{ \min_{u(k-1)} [S_{k-1}(f^{-1}(x(k) - Bu(k-1)) + [y(k) - g[x(k)]]'$$
$$\times [DLD']^{-1}[y(k) - g[x(k)]] + u'(k-1)K^{-1}u(k-1) \Big\} \qquad (7.58)$$

The procedure used to perform the minimization in (7.52) subject to (7.53), which led to the final result given by (7.58), is the "dynamic programming" formalism. Hence, it contains the usual computational difficulties inherent in the use of dynamic programming. The problem manifests itself here by the necessity to evaluate and store (for one iteration) the value of the function $S_k(x(k))$. For example, to evaluate $\hat{x}(k)$ in (7.58) we need the function S_{k-1}. With this knowledge every term in the right hand side is known except $x(k)$ and $u(k-1)$ which are subject to the minimization. The initial condition for the process is the function $S_0[x(0)]$ given by (from (7.55))

$$S_0[x(0)] = x'(0)P^{-1}(0)x(0)$$
$$+ [y(0) - g[x(0)]]'[DLD']^{-1}[y(0) - g[x(0)]] \qquad (7.59)$$

We can at least expect this dynamic programming method to work, although the practical computational difficulties seem to be very great.

PROBLEMS

1. A scalar random sequence $x(k)$ is defined by the solution of the following discrete dynamic system

$$x(k+1) = -0.5x(k) + 0.1 \sin x(k) + u(k)$$
$$Eu(k) = 0$$
$$Eu(k_1)u(k_2) = 0.1\Delta(k_2 - k_1)$$
$$x(0) = x_0$$
$$Ex_0 = 0$$
$$Ex_0^2 = 1$$

By means of a linearization procedure, derive an estimator for $x(k + 1)$ operating on $y(0), \ldots , y(k)$ where

$$y(k) = x(k) + 0.1x^3(k) + v(k)$$
$$Ev(k) = 0$$
$$Ev(k_1)v(k_2) = \Delta(k_2 - k_1)$$

2. Derive the continuous counterpart of equation (7.19).

3. Following a procedure similar to that used in Section 7.4, derive a recursive relationship for the evolution of the conditional probability density function

$$p[x(k) \mid y(0), \ldots , y(k)]$$

where the terms $x(k)$, $y(k)$ are defined in Section 7.2. Discuss the practical implications of your result in comparison with equation (7.29).

4. Let $x(k)$ and $y(k)$ be given by

$$x(k + 1) = x(k)$$
$$y(k) = x(k) + v(k)$$
$$Ev(k) = 0$$
$$Ev(k_1)v(k_2) = \Delta(k_2 - k_1)$$

where $v(k)$ is a gaussian (white noise) sequence. Show that the mode of the *a posteriori* density $p[x(1) \mid y(0), y(1)]$ coincides with the value of $x(1)$ in the pair $x(0)$, $x(1)$ maximizing the probability density

$$p[x(0), x(1) \mid y(0), y(1)]$$

5. Show that in the general linear gaussian case (i.e., when f and g in equations (7.1) and (7.3) are linear in $x(k)$ and when $u(k)$ and $v(k)$ are gaussian), the value of $x(k)$ maximizing $p[x(k) \mid y(0), \ldots , y(k)]$ is the same as the terminal term of the sequence $x(0), \ldots , x(k)$ maximizing $p[x(0), \ldots , x(k) \mid y(0), \ldots , y(k)]$.

SELECTED READINGS

Bass, R. W. *et al.*, "Optimal Multichannel Nonlinear Filtering," *J. Mathematical Analysis and Applications*, **16**, 1, 1966.

Bellman, R. E. and S. E. Dreyfus, *Applied Dynamic Programming*, Princeton University Press, Princeton, N.J., 1962.

Cox, H., "On the Estimation of State Variables and Parameters for Noisy Dynamic Systems," *IEEE Transactions on Automatic Control*," January 1964.

Detchemendy, D. M. and R. Sridhar, "Sequential Estimation of States and Parameters in Noisy Nonlinear Dynamical Systems," *J. Basic Engineering*, June 1966.

Ho, Y. C. and R. C. K. Lee, "A Bayesian Approach to Problems in Stochastic Estimation and Control," *IEEE Transactions on Automatic Control,* October 1964.

Kalman, R. E., "A New Approach to Linear Filtering and Prediction Problems," *J. Basic Engineering, ASME Transactions,* December 1960.

Kushner, H. J., "Nonlinear Filtering: The Exact Dynamical Equations Satisfied by the Conditional Mode," *IEEE Transactions on Automatic Control,* Vol. AC-12, June 1967.

Kushner, H. J., "On the Differential Equations Satisfied by Conditional Probability Densities of Markov Processes, With Applications," *J. SIAM Control,* 2, No. 1, 1962.

Wonham, W. M., "Some Applications of Stochastic Differential Equations to Optimal Nonlinear Filtering," *J. SIAM Control,* Ser. A., 2, No. 3, 1965.

chapter 8 / MAXIMUM LIKELIHOOD AND LEAST SQUARE ESTIMATION

8.1 Introduction

In the estimation problems considered in the preceding chapters, knowledge of the probability density function of the parameter to be estimated was always required prior to the reception of any observations. In many practical cases this information is not available. The parameter to be estimated may not even be a random variable. One approach to such problems is to interpret the lack of knowledge concerning the *a priori* probability density function of the parameter in the sense that the density function is implicitly assumed to be uniform (or approximately so over a very wide range). With this conjecture, the problem falls within the categories discussed before.

However, it is often desirable to use an estimation concept free of such assumptions. The estimation procedures satisfying this requirement are maximum likelihood, least square curve fitting, and method of moments. This chapter is devoted to study of these estimators, their interrelationships, and their connection with the results of previous chapters.

8.2 Maximum Likelihood Estimation

A reasonable estimate of a parameter is that value which will make a given observation most likely, i.e., the parameter value which causes the conditional probability density induced on the observations to have its greatest maximum at the given observation. This estimate is called the maximum likelihood estimate.

More precisely, let x be the parameter to be estimated (in general an n-vector). Let $y(i)$, $0 \leq i \leq k$, be a sequence of observations (each $y(i)$ in general is an s-dimensional vector) which are generated by the functional relationship

$$y(i) = f[x, v(i), i], \qquad i = 0, \ldots, k \qquad (8.1)$$

where each $v(i)$ in general is a q-dimensional vector representing observation noise or other random interference which tends to make it impossible to infer the true value of x from the observations $y(i)$. The probability density on $v(i)$ must be known so that, together with the functional relationship (8.1), it is possible to determine the conditional probability density $p[y(0), \ldots, y(k) \mid x]$ on the observations $y(i)$, $0 \leq i \leq k$, *given* the parameter value x. Naturally, if in a given problem the probability density $p[y(0), \ldots, y(k) \mid x]$ were known in advance, it would not be important to have information on the statistics of $v(i)$. Assuming now that the conditional probability density $p[y(0), \ldots, y(k) \mid x]$ is known, or has been derived from (8.1) and the statistics of $v(i)$, we may define the "likelihood function"

$$l(x) \triangleq p[y(0), \ldots, y(k) \mid x] \qquad (8.2)$$

where the conditional density $p[y(0), \ldots, y(k) \mid x]$ is now assumed to have been evaluated at a given received observation sequence, $y(0), \ldots, y(k)$.

The maximum likelihood estimate of x, denoted by $\hat{x}$, is now the value of x which maximizes $l(x)$ which was the conditional probability density function $p[y(0), \ldots, y(k) \mid x]$ evaluated for the given received observations. If (8.2) does not possess a unique relative maximum, then $\hat{x}$ is defined to correspond to the largest of its peaks. Notice that here we have not required knowledge of the probability density function $p(x)$. Instead, we have assumed that the conditional density $p[y(0), \ldots, y(k) \mid x]$ is known. Further, we have neither defined nor utilized any cost function (i.e., which would have associated a cost to each possible estimation error $x - \hat{x}$). Some properties of maximum likelihood estimators are discussed in the following.

We recall that the function $l(x)$ defined by (8.2) was called the likelihood function. Assuming that the required maximization of $l(x)$ can be achieved by differentiation, the estimation satisfies the equations (defined as likelihood equations)

$$\frac{\partial l(x)}{\partial x_1} = 0$$

$$\vdots \qquad (8.3)$$

$$\frac{\partial l(x)}{\partial x_n} = 0$$

or compactly

$$\text{grad } l(x) \triangleq \frac{\partial l(x)}{\partial x} = 0 \tag{8.4}$$

It is sometimes convenient, when $l(x)$ is an exponential function of x, to utilize the monotonicity of the natural logarithm to write an equivalent set of likelihood equations

$$\frac{\partial \log l(x)}{\partial x_1} = 0, \qquad \frac{\partial \log l(x)}{\partial x_2} = 0, \text{ etc.}$$

It can be shown, under rather general conditions on the form of the conditional density $p[y(0), \ldots, y(k) \mid x]$, that the maximum likelihood estimator tends to become unbiased as k (the number of observations in the observation sequence) approaches infinity.* In statistical parlance, we say that the maximum likelihood estimator is "asymptotically unbiased." This remarkable asymptotic property of maximum likelihood estimators does not hold, in general, for other types of estimators.

If we were to assume a uniform prior probability density function $p(x)$ for the parameter x, then from Bayes' formula we would have

$$p[x \mid y(0), \ldots, y(k)] = \frac{p[y(0), \ldots, y(k) \mid x]p(x)}{py((0), \ldots, y(k))} \tag{8.5}$$

Evaluating (8.5) at the received observation yields a number (independent of x) for the denominator in the right hand side. Since $p(x)$ is uniform, the value of x which will maximize the right hand side (i.e., the maximum likelihood estimate) coincides with the mode of the *a posteriori* density function $p[x \mid y(0), \ldots, y(k)]$.

▶ **example 1**
Let $y(i)$ be a scalar gaussian independent random sequence with unknown mean x_1 and variance x_2, i.e.,

$$Ey(i) = x_1$$

$$E[y(k_1) - x_1][y(k_2) - x_1] = x_2 \Delta(k_2 - k_1)$$

It is desired to estimate x_1 and x_2 by processing the observations $y(0), \ldots, y(k)$. The likelihood function is (from Eq. 8.2)

$$l(x) = \frac{1}{(2\pi x_2)^{(k+1)/2}} \exp -\frac{1}{2} \frac{\sum_{i=0}^{k} [y(i) - x_1]^2}{x_2}$$

* Cramer, H., *Mathematical Methods of Statistics.*

and the likelihood equations are given by (8.4) which become, after cancelling factors which are not zero,

$$\sum_{i=0}^{k} [y(i) - \hat{x}_1] = 0 \tag{8.6}$$

$$-(k + 1) + \frac{1}{x_2} \sum_{i=0}^{k} [y(i) - x_1]^2 = 0 \tag{8.7}$$

Solving (8.6) and (8.7) for x_1 and x_2 yields

$$\hat{x}_1 = \frac{1}{k + 1} \sum_{i=0}^{k} y(i) \tag{8.8}$$

$$\hat{x}_2 = \frac{\sum_{i=0}^{k} [y(i) - \hat{x}_1]^2}{k + 1} \tag{8.9}$$

Therefore the maximum likelihood estimates of the mean and variance are the sample mean and sample variance. It is interesting to compare this result with that of a similar example in Section 4.4 where we derived an estimator which minimized $E(x_1 - \hat{x}_1)^2$. The result then was an estimator which in the limit approached the sample mean.* However, the estimation of x_2 represented a nonlinear problem and could not have been solved by ▶ the methods discussed in Chapter 4.

▶ example 2

It is desired to derive the maximum likelihood estimate of the signal power, defined as x^2, in the signal

$$y(i) = xs(i), \qquad i = 0, 1, 2, \ldots, k$$

where $s(i)$ is a scalar, stationary, zero-mean gaussian random sequence with autocorrelation

$$Es(i)s(j) = \varphi(j - i)$$

Introducing the $(k + 1)$-vector s defined by

$$s \triangleq \begin{bmatrix} s(0) \\ s(1) \\ \cdot \\ \cdot \\ \cdot \\ s(k) \end{bmatrix}$$

* Either as k or $P(0)$ approaches infinity. The condition $P(0) = \infty$ is satisfied if we assume $p(x)$ to be uniform.

we observe that we have sufficient information to write the $(k + 1)$-dimensional gaussian probability density on the components of s:

$$p(s) = \frac{1}{(2\pi)^{(k+1)/2}|Q|^{1/2}}\exp\left\{-\tfrac{1}{2}s'Q^{-1}s\right\}$$

where

$$Q \triangleq \text{Cov } s = \begin{bmatrix} \varphi(0) & \varphi(1) & \varphi(2) & \cdots & \varphi(k) \\ \varphi(1) & \varphi(0) & \varphi(1) & \cdots & \varphi(k-1) \\ \varphi(2) & \varphi(1) & & & \vdots \\ \vdots & & \ddots & & \vdots \\ \vdots & & & \ddots & \vdots \\ \varphi(k) & \varphi(k-1) & & & \varphi(0) \end{bmatrix}$$

Introducing the $(k + 1)$-dimensional vector

$$y = \begin{bmatrix} y(0) \\ y(1) \\ y(2) \\ \vdots \\ \vdots \\ y(k) \end{bmatrix}$$

the conditional density $p(y \mid x)$, for each value of the scalar parameter x, is then given by

$$l(x) = p(y \mid x) = \frac{1}{(2\pi)^{(k+1)/2}|Q|^{1/2}(x^2)^{(k+1)/2}}\exp\left\{-\frac{1}{2x^2}y'Q^{-1}y\right\}$$

Regarding x^2 as the parameter we wish to estimate, the likelihood equation becomes

$$\frac{\partial l(x^2)}{\partial(x^2)} = \frac{(x^2)^{(-k+1)/2}}{(2\pi)^{(k+1)/2}|Q|^{1/2}}$$

$$\times\left[-\frac{(k+1)}{2}(x^2)^{-1} + \tfrac{1}{2}(x^2)^{-2}y'Q^{-1}y\right]\exp\left\{-\frac{1}{2x^2}y'Q^{-1}y\right\}$$

This yields, upon cancellation of various factors,

$$(\hat{x}^2) = \frac{y'Q^{-1}y}{k+1}$$

For the special case when $s(i)$ is an independent sequence with constant variance σ^2 (as in the previous example), the matrix Q becomes

$$Q = \sigma^2 I$$

and the determinant of Q becomes

$$|Q| = (\sigma^2)^n$$

and Q^{-1} becomes

$$Q^{-1} = \frac{1}{\sigma^2} I.$$

In this case, the estimator $(\hat{x}^2)$ becomes

$$(\hat{x}^2) = \frac{1}{\sigma^2} \frac{y^1 y}{(k+1)} = \frac{\sum\limits_{i=0}^{k} y^2(i)}{\sigma^2(k+1)}$$

8.3   Connection With Least Square Estimation

Let the observation noise $v(i)$ in (8.1) be additive and represent a zero-mean independent gaussian random sequence. In symbols, we have

$$y(i) = h(x, i) + v(i) \qquad i = 0, \ldots, k \tag{8.10}$$

$$Ev(i) = 0$$

$$Ev(i)v'(j) = L(j, i)\,\Delta(j - i) \tag{8.11}$$

where $L(i, i)$ is a sequence of known covariance matrices, each covariance matrix representing the covariance among the components of $v(i)$.

It is desired to determine the maximum likelihood estimate of the parameter x. The likelihood function, i.e., the conditional probability of $y(0), \ldots, y(k)$ given x is given by

$$l(x) = C \exp - \frac{1}{2} \sum_{i=0}^{k} [y(i) - h(x, i)]' L^{-1}(i, i)[y(i) - h(x, i)] \tag{8.12}$$

where C (which involves factors of $\sqrt{2\pi}$ and $|L(i, i)|$) is a constant independent of x. From the form of (8.12), we see that maximizing $l(x)$ leads to minimizing the quadratic function

$$\min_{x} \sum_{i=0}^{k} [y(i) - h(x, i)]' L^{-1}(i, i)[y(i) - h(x, i)] \tag{8.13}$$

If we interpret i as the time t_i and then consider the limit of t_i^* such that

* *See* Helstrom, C. W., *Statistical Theory of Signal Detection*.

the intervals $t_{i+1} - t_i$ approach zero and furthermore $t_k = T$, equation (8.13) leads to an "equivalent" continuous minimization

$$\min_x \int_0^T [y(t) - h(x, t)]' L^{-1}(t)[y(t) - h(x, t)] \, dt \qquad (8.14)$$

The operations specified by (8.13) and (8.14) define the method of weighted least square (in discrete and continuous form respectively) for estimating the parameter x when the observations consist of a discrete sequence, or a continuous function observed for $t \in [0, T]$. If $L(t)$ or $L(i, i)$ were identity matrices,* then we would have the ordinary least square problem. In the ordinary least square problem, we simply choose $\hat{x}$ such that the expected observation (i.e., $y(i) = h(x, i)$, which ignores the noise $v(i)$) comes as close as possible to the actual observation in least square sense of (8.13) or (8.14). It was shown above that the method of maximum likelihood and least square yield the same result in the special case of additive white gaussian noise. Notice that a stochastic optimization problem characterized by maximum likelihood estimation is in fact replaced by a deterministic optimization problem defined by (8.13) or (8.14).

▶ **example 1**
Let a scalar signal $h(x, t)$ be given by

$$h(x, t) = a \sin (t + \theta) \qquad (8.15)$$
with

$$x \triangleq \begin{bmatrix} a \\ \theta \end{bmatrix} \qquad (8.16)$$

where a and θ are unknown constant parameters which we should like to estimate.

Let the noise $v(t)$ be stationary additive white gaussian noise with zero mean and covariance function

$$L(t) = L_0 = \text{constant} \qquad (8.17)$$

so that we are dealing with the ordinary (nonweighted) least square problem.

It is desired to derive the maximum likelihood estimates of a and θ given the observation

$$y(t) = a \sin (t + \theta) + v(t); \qquad 0 \le t \le T \qquad (8.18)$$

This example will be discussed in detail since, in addition to providing an appropriate illustration of maximum likelihood estimation, it represents an important practical problem. Since the noise is white gaussian and

---

* Or differ from identity matrices by the same arbitrary multiplicative constant.

additive, we can use the result given by (8.14)

$$\min_{a,\theta} \int_0^T \frac{[y(t) - a \sin (t + \theta)]^2}{L_0} \, dt \tag{8.19}$$

To obtain the likelihood equations, we differentiate (8.19) under the integral sign and equate the derivatives to zero, yielding

$$\int_0^T [y(t) - \hat{a} \sin (t + \hat{\theta})] \sin (t + \hat{\theta}) \, dt = 0 \tag{8.20}$$

$$\int_0^T [y(t) - \hat{a} \sin (t + \hat{\theta})] \cos (t + \hat{\theta}) \, dt = 0 \tag{8.21}$$

Equations (8.20) and (8.21) represent a pair of algebraic equations which should be solved simultaneously for $\hat{a}$ and $\hat{\theta}$, the desired maximum likelihood estimates.

If only the amplitude a was to be estimated (i.e., $\theta = \theta_0$ was known), we would only need one likelihood equation, namely (8.20), which yields

$$\hat{a} = \frac{\int_0^T y(t) \sin (t + \theta_0) \, dt}{\int_0^T \sin^2 (t + \theta_0) \, dt} \tag{8.22}$$

A measure of the quality of this estimate can be obtained by evaluating the mean and variance of the conditional estimation error,* $a - \hat{a}$, i.e., the difference between the true value a and the estimate $\hat{a}$, given that the true value is a. From (8.18) and (8.22)

$$E(a - \hat{a}) = E\left\{ a - \frac{\int_0^T y(t) \sin (t + \theta_0) \, dt}{\int_0^T \sin^2 (t + \theta_0) \, dt} \right\} = E \frac{\int_0^T v(t) \sin (t + \theta_0) \, dt}{\int_0^T \sin^2 (t + \theta_0) \, dt} = 0 \tag{8.23}$$

Consequently (8.22) yields an unbiased estimator. The variance is given by

$$E(a - \hat{a})^2 = E\left\{ \frac{\int_0^T v(t) \sin (t + \theta_0) \, dt}{\int_0^T \sin^2 (t + \theta_0) \, dt} \right\}^2 \tag{8.24}$$

* Here $\hat{a}$ is a random variable being a function of the noisy observations $y(t)$, while a is a nonrandom constant.

Since $v(t)$ is white noise with covariance function $L_0 \delta(\tau)$ equation (8.24) can be further simplified yielding

$$E(a - \hat{a})^2 = \frac{L_0}{\displaystyle\int_0^T \sin^2 (t + \theta_0)\, dt} \tag{8.25}$$

It is observed in (8.25) that the variance is proportional to L_0 and tends to zero as T approaches infinity.

If the amplitude $a = a_0$ was known and it was desired to estimate θ, we would again need only one likelihood equation, namely (8.21), which yields

$$\int_0^T y(t) \cos (t + \theta)\, dt = -\frac{a_0}{4} [\cos 2(T + \theta) - \cos 2\theta] \tag{8.26}$$

Expanding by trigonometric identities, it follows that

$$\cos \theta \int_0^T y(t) \cos t\, dt - \sin \theta \int_0^T y(t) \sin t\, dt$$

$$= -\frac{a_0}{4} [\cos 2(T + \theta) - \cos 2\theta] \tag{8.27}$$

In the special case where T (the observation interval) is chosen to be an integer multiple of π the right hand side will equal zero. Consequently

$$\theta = \tan^{-1} \left\{ \frac{\displaystyle\int_0^T y(t) \cos t\, dt}{\displaystyle\int_0^T y(t) \sin t\, dt} \right\} \tag{8.28}$$

Notice that with the above restriction on T, the estimate $\hat{\theta}$ is independent of a_0. Consequently (8.31) along with (8.19) (when θ_0 is substituted by $\hat{\theta}$) constitute the solution to the problem of simultaneous maximum likelihood amplitude and phase estimation represented by equations (8.20) and (8.21). From (8.18) and (8.28) we can derive the mean and variance of $\theta - \hat{\theta}$ conditioned on θ (i.e., θ is an unknown constant and $\hat{\theta}$ is a random variable
▶ as a consequence of its dependence on $v(t)$.) This is left as an exercise.

▶ **example 2**
A target is flying on a straight line course with initial position x_1 and constant velocity x_2. The instantaneous position of the target is observed and contains white gaussian noise $v(t)$.

$$y(t) = x_1 + x_2 t + v(t) \qquad 0 \le t \le T \tag{8.29}$$

$$Ev(t_1)v(t_2) = l\delta(t_2 - t_1) \tag{8.30}$$

It is desired to derive the maximum likelihood estimates of x_1 and x_2, after the entire data record $y(t)$, $t \in [0, T]$, is obtained. Equation (8.14) becomes

$$\min_{x_1, x_2} \int_0^T [y(t) - x_1 - x_2 t]^2 \, dt \qquad . \qquad (8.31)$$

which physically means that we try to fit the observed data with a straight line in least square sense.

Differentiating (8.31) with respect to x_1 and x_2 and equating to zero yields

$$\int_0^T [y(t) - \hat{x}_1 - \hat{x}_2 t] \, dt = 0 \qquad (8.32)$$

$$\int_0^T t[y(t) - \hat{x}_1 - \hat{x}_2 t] \, dt = 0 \qquad (8.33)$$

Performing the integrals in (8.32) and (8.33) and solving for $\hat{x}_1$ and $\hat{x}_2$ results in

$$\hat{x}_1 = \frac{12}{T^3} \left\{ \tfrac{1}{3} T^2 \int_0^T y(t) \, dt - \tfrac{1}{2} T \int_0^T t y(t) \, dt \right\} \qquad (8.34)$$

$$\hat{x}_2 = \frac{12}{T^3} \left\{ -\tfrac{1}{2} T \int_0^T y(t) \, dt + \int_0^T t y(t) \, dt \right\} \qquad (8.35)$$

It can easily be shown that

$$E\hat{x}_1 = x_1$$

$$E\hat{x}_2 = x_2$$

$$E(x_1 - \hat{x}_1)^2 = 4 \frac{l}{T} \qquad (8.36)$$

$$E(x_2 - \hat{x}_2)^2 = \frac{12}{T^3} l$$

Hence the estimators are unbiased and the variances of the estimation errors tend to zero as T approaches infinity. It is suggested that the reader ▶ compare the results of this example with Example 2 of Section 4.9.

8.4 Recursive Least Square Estimation

The classical technique of least square estimation defined in the preceding section appears over and over again in a multitude of contemporary applications. Its attraction is the simplicity of the procedure as evidenced by the relationships (8.13) or (8.14) and the two preceding examples. Its shortcoming is that in general as more observations $y(t)$ are received, the complete operation has to be repeated, in addition to the fact that the entire past observations $y(t)$, $t \in [0, T]$, should be stored and utilized at each step.

However, when the observations are linear functions of the parameters to be estimated (as in Example 2), a recursive procedure may be developed which enables one to estimate the parameter value based only on the last estimate and the last additional observation, as follows:

Let us assume the observation $y(i)$ is a scalar.* Further, let us assume that the function $h(x, i)$ in (8.10) is linear in x, so that (8.10) can be written as

$$y(i) = g'(i)x + v(i) \qquad i = 0, \ldots, k \qquad (8.37)$$

where $g(i)$ and x are n-vectors and $y(i)$ and $v(i)$ are scalars. Let $v(i)$ be a zero-mean gaussian white noise sequence. Let us adopt the notation

$$y_k = \begin{bmatrix} y(0) \\ \cdot \\ \cdot \\ \cdot \\ y(k) \end{bmatrix}$$

$$G_k = \begin{bmatrix} g'(0) \\ \cdot \\ \cdot \\ \cdot \\ g'(k) \end{bmatrix} \qquad (8.38)$$

$$v_k = \begin{bmatrix} v(0) \\ \cdot \\ \cdot \\ \cdot \\ v(k) \end{bmatrix}$$

Hence y_k and v_k are $(k + 1)$-vectors and G_k is an $(k + 1) \times n$ matrix.

Using this notation, (8.37) becomes

$$y_k = G_k x + v_k \qquad (8.39)$$

with the condition that v_k is a vector-valued gaussian random variable whose components are independent

$$E v_k v'_k = L_k = \text{diag} (\sigma_0{}^2, \sigma_1{}^2, \ldots, \sigma_k{}^2) \qquad (8.40)$$

where $\sigma_0{}^2$, $\sigma_1{}^2$, etc., are the variances of the components of v_k.

* If $y(i)$ is vector-valued, each component can be considered a scalar observation and the problem remains the same.

The maximum likelihood estimate of x, given (8.39) and (8.40), is simply the solution to the following least square fitting problem: Choose x to minimize J_k defined by

$$J_k = [y_k - G_k x]' L_k^{-1} [y_k - G_k x]$$

Differentiating with respect to x and equating to zero yields

$$G_k' L_k^{-1} [y_k - G_k \hat{x}] = 0$$

Hence

$$\hat{x}_k = [G_k' L_k^{-1} G_k]^{-1} G_k' L_k^{-1} y_k \tag{8.41}$$

The notation $\hat{x}_k$ is used here to indicate that the last observation used to produce this estimate is $y(k)$.

Here it is assumed that the columns of G_k are linearly independent (i.e., the matrix G_k has rank n and hence the matrix $G_k' L_k^{-1} G_k$ is nonsingular since L_k is positive definite, because all $\sigma_0^2, \sigma_1^2, \ldots, \sigma_k^2$ are assumed nonzero). This assumption is usually satisfied in practice and is referred to as the observability condition. It implies that, in the absence of any noise we can solve for the unknown parameters exactly.* It is clear that if G_k has rank n for some k, additional observations which simply add more rows to the matrix G_k will not change its rank.

Let us define

$$P_k = [G_k' L_k^{-1} G_k]^{-1} \tag{8.42}$$

Hence (8.41) becomes

$$\hat{x}_k = P_k G_k' L_k^{-1} y_k \tag{8.43}$$

Now suppose an additional observation $y(k + 1)$ is received. The new estimate $\hat{x}_{k+1}$ then is similarly given by

$$\hat{x}_{k+1} = P_{k+1} G_{k+1}' L_{k+1}^{-1} y_{k+1} \tag{8.44}$$

The estimate $\hat{x}_{k+1}$ given by (8.44) can now be written as

$$\hat{x}_{k+1} = P_{k+1} \left[\begin{array}{c} G_k \\ \hline g'(k+1) \end{array} \right]' \left[\begin{array}{c|c} L_k & 0 \\ \hline 0 & \sigma_{k+1}^2 \end{array} \right]^{-1} \left[\begin{array}{c} y_k \\ \hline y(k+1) \end{array} \right] \tag{8.45}$$

From (8.43) and (8.45) we have

$$\hat{x}_{k+1} = \hat{x}_k + [P_{k+1} P_k^{-1} - I]\hat{x}_k + P_{k+1} g(k+1) \frac{y(k+1)}{\sigma_{k+1}^2} \tag{8.46}$$

where we have added and subtracted $\hat{x}_k$.

---

* This is very similar to the observability condition discussed in Chapter 3 where we derived the condition necessary for evaluating the unknown parameters (the initial condition for the differential equation) exactly in the absence of any input to the system.

The matrix $P(k + 1)$ can be written as follows

$$P_{k+1} = \left\{ \left[\begin{array}{c} G_k \\ \hline g'(k+1) \end{array} \right]' \left[\begin{array}{c|c} L_k & 0 \\ \hline 0 & \sigma^2_{k+1} \end{array} \right]^{-1} \left[\begin{array}{c} G_k \\ \hline g'(k+1) \end{array} \right] \right\}^{-1}$$

which becomes after performing the indicated matrix multiplications

$$P_{k+1} = \left\{ P_k^{-1} + \frac{g(k+1)g'(k+1)}{\sigma^2_{k+1}} \right\}^{-1} \tag{8.47}$$

Substituting for P_k^{-1} from (8.45) yields

$$\hat{x}_{k+1} = \hat{x}_k + P_{k+1}g(k+1) \frac{[y(k+1) - g'(k+1)\hat{x}_k]}{\sigma^2_{k+1}} \tag{8.48}$$

Now, it remains to express $P(k + 1)$ as a function of $P(k)$, in a convenient form. Equation (8.47) shows that P_{k+1} can be written as a function of P_k if we are willing to do the indicated $n \times n$ inversions. Fortunately, the matrix inversions can be avoided by using an identity which is well known in linear algebra. We have

$$P_{k+1} = P_k - P_k g(k+1)[g'(k+1)P_k g_k(k+1) + \sigma^2_{k+1}]^{-1} g'(k+1)P_k \tag{8.49}$$

where the right hand sides of (8.47) and (8.49) are equal to each other identically. This identity is verified by substituting for P_{k+1}^{-1} and P_{k+1} from (8.47) and (8.49), respectively, into the following

$$P_{k+1}^{-1} P_{k+1} = I + g(k+1) \left\{ \frac{1}{\sigma^2_{k+1}} g'(k+1)P_k g(k+1) \right.$$
$$\times [g'(k+1)P_k g(k+1) + \sigma^2_{k+1}]^{-1}$$
$$\left. - [g'(k+1)P_k g(k+1) + \sigma^2_{k+1}]^{-1} \right\} g'(k+1)P_k = I \tag{8.50}$$

The advantage of (8.49) over (8.47) is that the quantity in the bracket is a scalar, and hence no matrix inversion is required. We may now see that the equations (8.48) and (8.49) specify a very simple recursive solution to the least square estimation problem introduced in this section. In order to solve (8.49) for P_k we need an initial condition. This is obtained from (8.42) for some value of k for which the matrix $G_k' L_k^{-1} G_k$ is non-singular. Since G_k was assumed to have independent columns the matrix G_k and hence $G_k' L_k^{-1} G_k$ is nonsingular for $k = n - 1$.

Interpretation of the matrix P_k defined by (8.42) in the light of maximum likelihood estimation reveals an interesting result. Notice that least square estimation of a parameter is purely a deterministic problem which, incidentally, leads to the maximum likelihood estimate of the parameter if the only random noise in the system, namely, the observation noise, is zero-mean additive white gaussian.

As a maximum likelihood estimate we can determine the covariance of the estimation error $x - \hat{x}$ where $\hat{x}$ is given by the least squares procedure (8.43) and where the observations y_k are generated as in (8.39). Notice that here x is a nonrandom parameter (regarded as an unknown constant vector), whereas $\hat{x}$ is a random variable (due to its dependence on v_k). Substituting (8.39) into (8.43) yields

$$\hat{x}_k = P_k G_k' L_k^{-1} [G_k x + v_k] \tag{8.51}$$

Using the fact that v_k has zero-mean, and by the definition of P_k given in (8.42), we have

$$E\hat{x}_k = x \tag{8.52}$$

so that the maximum likelihood estimate $\hat{x}_k$ is unbiased. We may now evaluate the covariance matrix of the estimator error $x - \hat{x}_k$. From (8.51), we have

$$x - \hat{x}_k = -P_k G_k' L_k^{-1} v_k$$

from which it follows that

$$\mathrm{Cov}\,(x - \hat{x}_k) = E(x - \hat{x}_k)(x - \hat{x}_k)' = E P_k G_k' L_k^{-1} v_k v_k' (L_k^{-1})' G_k P_k' = P_k \tag{8.53}$$

▶ **example**

Consider a discrete version of Example 2 of the preceding section. Assuming that the observations are received at one-second intervals and that the noise has unity covariance function we have

$$y(i) = x_1 + x_2 i + v(i) \qquad 0 \le i \le k$$
$$Ev(i)v(j) = \Delta(j - i)$$

Consequently

$$g(k) = \begin{bmatrix} 1 \\ k \end{bmatrix}$$

so that the matrix G_k defined by (8.38) has the form

$$G_k = \begin{bmatrix} 1 & 0 \\ 1 & 1 \\ 1 & 2 \\ & \cdot \\ & \cdot \\ & \cdot \\ 1 & k \end{bmatrix}$$

We see that the columns of G_k will be independent for $k \geq 1$, and consequently the observability condition is satisfied (i.e., $G'_k L^{-1} G_k$ is nonsingular) if we make at least two observations.

The recursive least square estimator is then given by (8.48) and (8.49). The only remaining step is to specify a starting points (initial conditions) for (8.48) and (8.49). Since the observability condition is valid only for $k \geq 1$, initial conditions P_k and $\hat{x}_k$, $k = 1$ should be determined. The definition of P_k given by (8.42) leads to

$$
P_1 = \left\{ \begin{bmatrix} 1 & 1 \\ 0 & 1 \end{bmatrix} \begin{bmatrix} 1 & 0 \\ 1 & 1 \end{bmatrix} \right\}^{-1} = \begin{bmatrix} 1 & -1 \\ -1 & 2 \end{bmatrix}
$$

Hence, using (8.41) we also get

$$
\hat{x}_1 = \begin{bmatrix} y(0) \\ y(1) - y(0) \end{bmatrix}
$$

As the additional data $y(2)$, $y(3)$, ..., become available, the estimates ▶ $\hat{x}_2$, $\hat{x}_3$, ..., may be computed recursively by using (8.48) and (8.49).

8.5   Connection With Kalman Filtering

In this section, the connection between the discrete recursive least square estimator given by (8.48) and (8.49) and the Kalman discrete filter derived in Section 4.4 will be discussed.

Recall that in equation (8.37) the vector $g(i)$ was arbitrary. The only condition imposed on $g(i)$ later in the section was to require the matrix $G'_k L_k^{-1} G_k$ to be nonsingular.

Now let us consider the following estimation problem. Suppose we are given the n-dimensional difference equation

$$
x(k + 1) = Ax(k) \tag{8.54}
$$

where the initial condition is $x(0) = x$. Here, x will represent a fixed, but unknown parameter vector whose value we shall attempt to estimate. The matrix A in (8.54) is an $n \times n$ constant matrix.* Suppose we are given a sequence of scalar observations defined by

$$
y(i) = Cx(i) + v(i) \qquad 0 \leq i \leq k \tag{8.55}
$$

where $v(i)$ is scalar white gaussian noise sequence with $Ev^2(i) = L$.

It is desired to derive the least square (or maximum likelihood) estimate

---

* The assumption that A be a constant matrix is not necessary. It is merely made here to simplify the notation. Moreover, C in (8.55) may be a $q \times n$ matrix so that the observations $y(i)$ become vector-valued.

of $x(k + 1)$ by processing $y(i)$: $0 \leq i \leq k$. Solving (8.54) for $x(i)$ and substituting into (8.55) yields

$$y(i) = CA^i x + v(i) \qquad A^0 \triangleq I \tag{8.56}$$

Let us denote the least square estimates of $x(k + 1)$ and $x(0) = x$ $\big($given the observations $y(i)$, $i = 0, \ldots, k\big)$ by $\hat{x}(k + 1)$ and $\hat{x}_k$ respectively. By definition of the least square estimate of x using the observation (8.56) we have

$$\hat{x}_k = \min_x \sum_{i=0}^{k} [y(i) - CA^i x]^2$$

Furthermore, if A is nonsingular by substituting $A^{-k-1}x(k + 1)$ for x it follows that

$$\hat{x}(k + 1) = \min_{x(k+1)} \sum_{i=1}^{k} [y(i) - CA^{i-k-1}x(k + 1)]^2$$

But since

$$\min_x \sum_{i=0}^{k} [y(i) - CA^i x]^2 = \min_{\substack{A^{k+1} \\ x = x(k+1)}} \sum_{i=0}^{k} [y(i) - CA^i x]^2$$

Hence

$$\hat{x}(k + 1) = A^{k+1}\hat{x}_k \tag{8.57}$$

Consequently, we can concentrate on deriving the estimate $\hat{x}_k$. This is helpful since we have already obtained procedures for estimating a constant (such as x) while $x(k + 1)$ is clearly not a constant in general.

Comparison of (8.56) and (8.37) reveals that this equation is in the form of the one used in the problem solved in the preceding section if we identify $g(i)$ by

$$g'(i) = CA^i \qquad \text{or} \qquad g(i) = A^{i'}C' \tag{8.58}$$

The estimate $\hat{x}_k$ for x is then given by the recursive relations (8.47) and (8.49) after substitution from (8.56). Carrying out this substitution and furthermore substituting $k - 1$ for k we obtain

$$\hat{x}_k = \hat{x}_{k-1} + P_k A^{k'} C' L^{-1}(k)[y(k) - CA^k \hat{x}_{k-1}] \tag{8.59}$$

$$P_k = P_{k-1} - P_{k-1} A^{k'} C' [CA^k P_{k-1} A^{k'} C' + L(k)]^{-1} CA^k P_{k-1} \tag{8.60}$$

Let us denote the covariance of $x(k) - \hat{x}(k)$ by $P(k)$ and that of $x(k + 1) - \hat{x}(k + 1)$ by $P(k + 1)$. Consequently, from (8.54), (8.57)

$$P(k + 1) = E[x(k + 1) - \hat{x}(k + 1)][x(k + 1) - \hat{x}(k + 1)]'$$
$$= A^{k+1} P_k A'^{(k+1)} \tag{8.61}$$

Substituting for P_k and P_{k-1} from (8.61) into (8.60) yields

$$P(k + 1) = AP(k)A' - AP(k)C'[CP(k)C' + L^{-1}]CP(k)A \tag{8.62}$$

Substituting for $\hat{x}_k$ in (8.48) the value of $[A^{k+1}]^{-1}\hat{x}(k+1)$ from (8.57) and after some matrix manipulation we get

$$\hat{x}(k+1) = A\hat{x}(k) + F(k)[y(k) - C\hat{x}(k)] \qquad (8.63)$$

where

$$F(k) = AP(k)C'[CP(k)C' + L]^{-1}$$

Equations (8.62) and (8.63) are identical with (4.24) and (4.26) for the specific problem under consideration, namely when $K = 0$ (i.e., when the only random source in the system is the observation noise). Had we assumed the following equation

$$x(k+1) = Ax(k) + Bu(k)$$

$$Eu(k_1)u'(k_2) = K\,\Delta(k_2 - k_1)$$

in place of (8.54), the addition of the missing term BKB' to (8.62) could be justified, thereby rendering (8.62) and (8.63) completely equivalent to the equations of Kalman discrete estimator (one-step prediction).

The assumption that the matrix $G_k'L_k^{-1}G_k$ (for $k \geq n - 1$) be nonsingular is clearly satisfied if $G_k'G_k$ is nonsingular since L_k is positive definite. (Note that $G_k'G_k$ to be nonsingular implies that the columns of G_k are linearly independent, since $G_k'G_k$ is the Gramian matrix of G_k.) The statement that $G_k'G_k$ is nonsingular is equivalent to the following matrix being positive definite*

$$\sum_{i=0}^{n-1} A^{i'}C'CA^i > 0$$

which is the discrete version of the observability condition given by equation (3.35).

A brief discussion of the results of this section is warranted. First, observe that the results of least square estimation (a purely deterministic operation) with a probabilistic interpretation (via maximum likelihood) agrees in form with the Kalman linear estimator minimizing average quadratic cost.

Second, the solution to the Kalman estimation equations requires knowledge of the initial value of covariance matrix. This is the same as requiring a prior density function for the parameter to be estimated. Maximum likelihood estimation does not require such data, and consequently the initial conditions for equations (8.47) and (8.49) are not given *a priori*. Instead we can wait until n observations are received and then start equation (8.49) with

---

* The matrix $G_k'G_k$ being nonsingular for $k = n - 1$ implies that the rows of G_k are linearly independent. These rows are $C, CA, \ldots, CA^{n-1}$. Consequently, there exists no nonzero n-dimensional vector z such that $CA^iz = 0$, $i = 0, \ldots, n - 1$. Hence at least one of the terms $z'A^{i'}C'CA^iz = \|CA^iz\|^2$; $i = 0, \ldots, n - 1$ is nonzero. Since no such term can be negative, it follows that $\sum_{i=0}^{n-1} A^{i'}C'CA^i$ is positive definite.

the condition

$$P_{k-1} = [G'_{k-1} L^{-1}_{k-1} G_{k-1}]^{-1}; \qquad k = n$$

and equation (8.48) with

$$\hat{x}_{k-1} = P_{k-1} G'_{k-1} L^{-1}_{k-1} Y_{k-1}$$

which means that we have to perform one matrix inversion after n observations are received and then proceed with equations (8.47) and (8.49). In other words, since we do not have the *a priori* density function required, we will wait for the first opportunity before establishing a probability density functions for the parameters. Since there are n parameters involved (the n components of x) we need at least n observations. As a consequence of the condition of observability n observations are also sufficient for the purpose. The example of the preceding section can be used as an illustration of the point discussed above.

8.6   Curve Fitting

The deterministic procedure of least square estimation, or matching the observed data with a set of known functions in the "least square" sense of (8.13) or (8.14), was one of the earliest techniques for evaluation of unknown parameters. It still is a common and effective procedure, as a result of the inherent simplicity of the technique. In many practical applications, such as in cases when the observation noise is not white gaussian or when no statistical knowledge of this quantity is available, the weighting matrix $(L(i, i)$ in (8.13), and $L(t)$ in (8.14)) is often chosen arbitrarily (often the identity matrix is chosen, for simplicity). Of course, in such a case it is not expected that the least square estimate of the unknown parameters will be the same as their maximum likelihood estimates.*

The terminology "curve-fitting" has also been used in conjunction with least square estimation procedure. It is a descriptive title since actually we are fitting (in certain sense) a combination of known functions (curves) to the observed function. A natural and common choice for the functions is the linear combination of successive powers in t (or i). More specifically, given the observation $y(t)$; $0 \leq t \leq T$ we would like to determine the parameters $x_0, \ldots, x_n$ satisfying the following

$$\min_{x_0, \ldots, x_n} \int_0^T [y(t) - x_0 - x_1 t - \cdots - x_n t^n]^2 \, dt \qquad (8.64)$$

* In fact if no statistical information concerning the observation noise is available, the maximum likelihood estimate does not exist, since the conditional probability density on the observations (given the parameter value) does not exist.

or in discrete version

$$\min_{x_0, \ldots, x_n} \sum_{i=0}^{k} [y(i) - x_0 - x_1 i - \cdots - x_n i^n]^2 \qquad (8.65)$$

The operations suggested by (8.64) and (8.65) are referred to as polynomial fitting. Since the polynomial $x_0 + x_1 i + \cdots + x_n i^n$ is linear in the parameters $x_0, \ldots, x_n$ the recursive procedure developed in Section 8.4 is clearly applicable. The G_k matrix for $k = n - 1$ becomes

$$G_k = \begin{bmatrix} 1 & 0 & 0 & \cdots & 0 \\ 1 & 1 & 1 & \cdots & 1 \\ 1 & 2^1 & 2^2 & \cdots & 2^n \\ \cdot & & & & \\ \cdot & & & & \\ \cdot & & & & \\ 1 & (k-1)^1 & & & (k-1)^n \end{bmatrix}$$

We see that all of the columns of G_k are linearly independent, hence the observability condition is always satisfied. In a number of applications a more suitable choice for the "fitting" curve is a transcendental function such as $x_1 \sin(x_2 t + x_3)$ or combination of such functions. Example 1 of Section 8.3 can be used for illustration. Clearly since the observation

$$y(t) = x_1 \sin(x_2 t + x_3) + v(t)$$

is not linear in x_2 and x_3 the recursive procedure of Section 8.4 is not applicable here.

8.7 Maximum Likelihood Estimation with Colored Observation Noise

Let a sequence of scalar observations be given by

$$y(i) = h(x, i) + v(i) \qquad i = 0, \ldots, k \qquad (8.66)$$

where the observation noise sequence is zero-mean gaussian, but where the successive terms $v(i)$ are correlated, i.e.,

$$Ev(i)v(j) = l(i, j) \qquad (8.67)$$

The likelihood function follows as

$$C \exp - \frac{1}{2} \sum_{j=0}^{k} \sum_{i=0}^{k} [y(i) - h(x, i)] r(i, j)[y(j) - h(x, j)] \qquad (8.68)$$

where C is a constant independent of x and the matrix $[r(i, j)]$ is inverse of the covariance matrix $[l(i, j)]$.

Consequently

$$\sum_{j=0}^{k} l(i, j) r(j, m) = \Delta(i - m) \tag{8.69}$$

The maximum likelihood estimate of x then satisfies the following

$$\min_{x} \sum_{j=0}^{k} \sum_{i=0}^{k} [y(i) - h(x, i)] r(i, j)[y(j) - h(x, j)] \tag{8.70}$$

As in Section 8.3, if we identify i by t_i and let $t_{i+1} - t_i$ approach zero, the continuous version of (8.70) will assume the following form

$$\min_{x} \int_0^T \int_0^T [y(t_1) - h(x, t_1)] r(t_1, t_2)[y(t_2) - h(x, t_2)] \, dt_1 \, dt_2 \tag{8.71}$$

where the matrix $[r(t_1, t_2)]$ is the inverse of $[l(t_1, t_2)]$. Equation (8.69) in the limit takes the form of

$$\int_0^T l(t_1, t) r(t, t_2) \, dt = \delta(t_1 - t_2) \qquad 0 < t_1, \ t_2 < T \tag{8.72}$$

The likelihood equation corresponding to (8.71) becomes

$$\int_0^T \int_0^T \frac{\partial h(\hat{x}, t_1)}{\partial x} r(t_1, t_2)[y(t_2) - h(\hat{x}, t_2)] \, dt_1 \, dt_2 = 0 \tag{8.73}$$

which is equivalent to

$$\int_0^T \frac{\partial q(\hat{x}, t_2)}{\partial x} [y(t_2) - h(\hat{x}, t_2)] \, dt_2 = 0 \tag{8.74}$$

where $q(x, t_1)$ is the solution of the integral equation

$$\int_0^T l(t, t_1) q(\hat{x}, t) \, dt = h(\hat{x}, t_1) \tag{8.75}$$

▶ **example**

It is desired to estimate the amplitude x in the signal $xh(t)$ by processing the observed data $y(t)$ where

$$y(t) = xh(t) + v(t)$$
$$Ev(t_1)v(t_2) = \varphi(t_2 - t_1)$$

From (8.75) and since $xh(t)$ is linear in x we may define $q_1(t)$ as

$$\frac{\partial q(x, t)}{\partial x} = q_1(t)$$

where

$$\int_0^T \varphi(t_1 - t)q_1(t)\, dt = h(t_1) \tag{8.76}$$

and from (8.74)

$$\int_0^T q_1(t)[y(t) - \hat{x}h(t)]\, dt = 0$$

with the maximum likelihood estimate given by

$$\hat{x} = \frac{\displaystyle\int_0^T y(t)q_1(t)\, dt}{\displaystyle\int_0^T h(t)q_1(t)\, dt} \tag{8.77}$$

If $v(t)$ was white noise $\big($i.e., $\varphi(t_1 - t) = \delta(t_1 - t)\big)$ then from (8.76)

$$q_1(t) = h(t)$$

and (8.77) becomes

$$\hat{x} = \frac{\displaystyle\int_0^T y(t)h(t)\, dt}{\displaystyle\int_0^T h^2(t)\, dt} \tag{8.78}$$

Substituting for $y(t)$ in (8.77) yields

$$\hat{x}(t) = \frac{x\displaystyle\int_0^T h(t)q_1(t)\, dt + \int_0^T v(t)q_1(t)\, dt}{\displaystyle\int_0^T h(t)q_1(t)\, dt}$$

Consequently the estimator is unbiased since

$$E\hat{x}(t) = x$$

The variance of the estimator is given by

$$E[\hat{x} - x]^2 = \frac{\displaystyle\int_0^T \int_0^T \varphi(t_1 - t_2)q_1(t_1)q_1(t_2)\, dt_1\, dt_2}{\left[\displaystyle\int_0^T h(t)q_1(t)\, dt\right]^2} \tag{8.79}$$

Substituting (8.76) into the numerator of (8.79) yields

$$\blacktriangleright \qquad E[\hat{x} - x]^2 = \left[\int_0^T h(t)q_1(t)\, dt\right]^{-1}$$

8.8 Least Square Estimates of Solutions
of Nonlinear Dynamic Systems

Let us formulate the following problem: A system is described by the n-dimensional vector differential equation

$$\dot{x} = f(x, t)$$
$$x(t) = \text{unknown} \tag{8.80}$$

and the r-dimensional observation vector $y(\tau)$ is given by

$$y(\tau) = h[x(\tau), \tau] + v(\tau); \qquad 0 \leq \tau \leq t \tag{8.81}$$

where $v(\tau)$ is the observation noise. We would like to derive the least square estimate $\hat{x}(t)$ which minimizes the deterministic curve-fitting criterion function

$$J[x(t), t] \triangleq \int_0^t [y(\tau) - h(x, \tau)]'Q[y(\tau) - h(x, \tau)] \, d\tau \tag{8.82}$$

where Q is a given positive definite matrix.* In (8.82) it is understood that for each t, the data record $y(\tau)$, $0 \leq \tau \leq t$, is given, so that we have the purely deterministic problem of finding the function $\hat{x}(\tau)$, $0 \leq \tau \leq t$, which minimizes (8.82), taking into account the fact that $x(t)$ is the solution of the *known* differential equation (8.80).

A discrete version of this problem was solved in the special case where (8.80) represents a linear system in Section 8.5. According to the results of Section 8.3, the least square estimate $\hat{x}(t)$ to be obtained here will also be the maximum likelihood estimate of $x(t)$ if the observation noise $v(t)$ is zero-mean (additive) white gaussian.

The integral in (8.82) is clearly a function of t and $x(t)$, since t appears as an upper bound on the integral, and since $x(\tau)$ for $0 \leq \tau \leq t$ in the integrand depends on $x(t)$.†

Since $y(\tau)$ is a known, observed function, and $J[x(t), t]$ is a function of $x(t)$ the optimal estimate $\hat{x}(t)$ then satisfies the equation

$$J_x[\hat{x}(t), t] \triangleq \frac{\partial J[\hat{x}(t), t]}{\partial x} = 0 \tag{8.83}$$

or taking the total derivative of $J_x[\hat{x}(t), t]$, considering that it is equal to a constant (zero), yields

$$J_{xx}[\hat{x}(t), t] \, d\hat{x} + J_{xt}[\hat{x}(t), t] \, dt = 0 \tag{8.84}$$

---

* If we desire $\hat{x}(t)$ to be the maximum likelihood estimate of $x(t)$ and if $v(t)$ is white gaussian noise with covariance L then we should choose $Q = L^{-1}$.

† Of course, the usual assumption that (8.80) possesses a unique solution is necessary.

where J_{xx} is an $n \times n$ matrix with components $\dfrac{\partial^2 J}{\partial x_i\, \partial x_j}$ and J_{xt} is an n-vector with components $\dfrac{\partial^2 J}{\partial x_i\, \partial t}$. From (8.84) we have

$$\frac{d\hat{x}(t)}{dt} = -J_{xx}^{-1}[\hat{x}(t),\, t]J_{xt}[\hat{x}(t),\, t]. \tag{8.85}$$

This is an equation for the optimal estimate $\hat{x}(t)$ if we can evaluate the terms J_{xx} and J_{xt} in the right hand side. Differentiating (8.82) with respect to t yields

$$J_t + J_x'\dot{x} = [y(t) - h(x,\, t)]'Q[y(t) - h(x,\, t)] \tag{8.86}$$

Substituting for $\dot{x}$ from (8.80) and differentiating with respect to x results in

$$J_{tx} + J_{xx}f(x,\, t) + f_x'(x,\, t)J_x = -2h_x'Q[y(t) - h(x,\, t)]. \tag{8.87}$$

Multiplying both sides by J_{xx}^{-1} yields

$$J_{xx}^{-1}J_{tx} = -f(x,\, t) - 2J_{xx}^{-1}h_x'Q[y(t) - h(x,\, t)] - J_{xx}^{-1}f_x'(x,\, t)J_x. \tag{8.88}$$

The last term in (8.88) is zero when evaluated at $\hat{x}(t)$ (equation 8.83).

Substituting (8.88) into (8.85) the estimate equation follows as

$$\frac{d\hat{x}(t)}{dt} = f(\hat{x},\, t) + P(t)h_x'[\hat{x}(t),\, t]Q[y(t) - h(\hat{x}(t),\, t)] \tag{8.89}$$

where $P(t)$ is given by

$$P(t) = 2J_{xx}^{-1}[\hat{x}(t),\, t] \tag{8.90}$$

An equation for $P(t)$ is derived in the following. From (8.90)

$$J_{xx}\frac{dP(t)}{dt} + \left(\frac{d}{dt}J_{xx}\right)P(t) = 0 \tag{8.91}$$

or

$$\frac{dP(t)}{dt} = -\tfrac{1}{2}P(t)\left(\frac{d}{dt}J_{xx}\right)P(t) \tag{8.92}$$

where

$$\frac{d}{dt}J_{xx}[\hat{x}(t),\, t] = J_{xxt}[\hat{x}(t),\, t] + J_{xxx}[\hat{x}(t),\, t]\frac{d\hat{x}}{dt} \tag{8.93}$$

Since in many practical applications J can be well approximated by the terms in its Taylor series expansion up to second powers in x in the vicinity of $\hat{x}$, the second term in the right hand side of (8.93) can be ignored. The term J_{xxt} is obtained by differentiating (8.87) with respect to x. Again setting $J_{xxx} = 0$ and using (8.83) it follows that

$$J_{xxt} = -J_{xx}f_x - f_x'J_{xx} - 2h_{xx}'Q[y - h] + 2h_x'Qh_x \tag{8.94}$$

Substituting into (8.92) yields

$$\frac{dP(t)}{dt} = f_x P(t) + P(t)f'_x + P(t)\sum_{i=1}^{r}[(h'_x Q)^i_x(y_i - h_i)]P(t) - P(t)h'_x Q h_x P(t)$$
(8.95)

where the superscript i indicates the ith column of the matrix. The initial value $P(0)$ is given by

$$P(0) = 2J_{xx!}^{-1}[\hat{x}(0), 0]$$
(8.96)

Equations (8.89) and (8.95) define the least square estimator. These equations must be solved on-line in real time. Unlike the linear case the covariance equation (8.95) depends on $\hat{x}(t)$ and the observation y, and consequently it cannot be solved before the observations are received. Finally, an interesting observation is the great similarity between equations (8.89) and (8.95) and those derived in Section 7.3 of the preceding chapter.

▶ example
Let

$$f(x, t) = Ax$$
$$h(x, t) = Cx$$

where A and C are $n \times n$ and $r \times n$ constant matrices. The observation noise is white gaussian with covariance function L. Consequently

$$Q = L^{-1}$$
$$h_x = C$$
$$f_x = A$$
$$h_{xx} = 0$$

Hence (8.89) and (8.95) become

$$\frac{d\hat{x}(t)}{dt} = A\hat{x}(t) + P(t)C'L^{-1}[y(t) - C\hat{x}(t)]$$
(8.97)

$$\frac{dP(t)}{dt} = AP(t) + P(t)A' - P(t)C'L^{-1}CP(t)$$
(8.98)

Comparison of (8.97) and (8.98) with (4.50) and (4.51) reveals that the estimator equations are identical with those of the Kalman-Bucy estimator for the special case under consideration namely

▶ $$B = K = 0, \qquad D = I$$

8.9   Method of Moments

The method of moments is another procedure for providing an estimate of a parameter without requiring the *a priori* knowledge of its probability density function although, as in the case of maximum likelihood estimation, a conditional probability density on the observations is required. It yields an estimate which is not necessarily optimal in any sense. Yet, the method is intuitively appealing due to its simplicity. In many cases, the estimate approaches the true value of the parameter as the interval of observation becomes infinite, or as the amount of observed data becomes infinite.

Let the parameter to be estimated be the n-vector x, and let the observation be given by the scalar random variable $y(i)$, $0 \leq i \leq k$, which is assumed to have a known conditional probability density function $p[y(i) \mid x]$. Using this density function, an expression for the first n moments* of $y(i)$, denoted by m_j, $1 \leq j \leq n$, can be derived in terms of the parameter x:

$$m_j(x) = \int [y(i)]^j p[y(i) \mid x] \, dy(i) \qquad j = 1, \ldots, n \qquad (8.99)$$

In (8.99) it is assumed that the successive observations $y(i)$ are independent and that $p[y(i) \mid x]$ is the same for all i. Now, the first n sample moments, denoted by μ_j, $1 \leq j \leq n$ can be evaluated from the observations $y(i)$, $i = 1, \ldots, k$, namely,

$$\mu_j = \frac{1}{k} \sum_{i=1}^{k} y^j(i) \qquad i \leq j \leq n \qquad (8.100)$$

Finally, the estimate of x is chosen to satisfy the n simultaneous equations given by

$$m_j(x) = \mu_j \qquad j = 1, \ldots, n \qquad (8.101)$$

Note that the number of observations $y(i)$ need not be equal to or greater than the number of parameters to be estimated. There need only be enough observations so that the n-moment equations (8.101) have a nontrivial solution. However, in general, the estimation error is a decreasing function of k.

▶ example

Let $y(i)$ be a scalar gaussian random variable with unknown mean denoted by x_1 and unknown variance denoted by x_2. We would like to estimate x_1 and x_2 by processing the observed data $y(i)$, $0 \leq i \leq k$, where $y(i)$ is an independent sequence.

Since there are two unknowns, two simultaneous moment equations are

---

* Here, the first n moments are used for simplicity. In general, any set of n moments could be used.

necessary. The first two moments of $y(i)$ are given by

$$m_1(x_1, x_2) = \int \frac{y(i)}{\sqrt{2\pi x_2}} \exp - \frac{1}{2} \frac{[y(i) - x_1]^2}{x_2} \, dy(i)$$

$$= x_1 \tag{8.102}$$

$$m_2(x_1, x_2) = \int \frac{y^2(i)}{\sqrt{2\pi x_2}} \exp - \frac{1}{2} \frac{[y(i) - x_1]^2}{x_2} \, dy(i)$$

$$= x_2 + x_1^2 \tag{8.103}$$

Therefore, we have the following simultaneous equations in terms of the estimates $\hat{x}_1$ and $\hat{x}_2$

$$\hat{x}_1 = \frac{1}{k} \sum_{i=1}^{k} y(i)$$

$$\hat{x}_1^2 + \hat{x}_2 = \frac{1}{k} \sum_{i=1}^{k} y^2(i) \tag{8.104}$$

Hence

$$\hat{x}_2 = \frac{1}{k} \sum_{i=1}^{k} y^2(i) - \left[\frac{1}{k} \sum_{i=1}^{k} y(i)\right]^2 = \frac{1}{k} \sum_{i=1}^{k} \left[y(i) - \frac{1}{k} \sum_{j=1}^{k} y(j)\right]^2 \tag{8.105}$$

The quality of these estimates can be obtained by deriving the means and the variances of $\hat{x}_1$ and $\hat{x}_2$. Recall that x_1 and x_2 are deterministic quantities while $\hat{x}_1$ and $\hat{x}_2$ are random variables because of their dependence on $y(i)$.

From (8.104) the expected value of $\hat{x}_1$ is given by

$$E\hat{x}_1 = \frac{1}{k} \sum_{i=1}^{k} Ey(i) = x_1$$

Hence the estimator for x_1 is unbiased. The variance is given by the expected value of $[x_1 - \hat{x}_1]^2$, i.e.,

$$E[x_1 - \hat{x}_1]^2 = \frac{1}{k^2} \sum_{j=1}^{k} \sum_{j=1}^{k} E[y(i) - x_1][y(j) - x_1] \tag{8.106}$$

If the random sequence $y(i)$, $i = 1, \ldots, k$ was uncorrelated, then (8.106) reduces to

$$E[x_1 - \hat{x}_1]^2 = \frac{1}{k} \sum_{i=1}^{k} E[y(i) - x_1]^2 = \frac{x_2}{k} \tag{8.107}$$

It is seen that as k approaches infinity the variance of the estimation error $x_1 - \hat{x}_1$ tends to zero. In the same manner, the expected value of $x_2 - \hat{x}_2$ can be obtained yielding

$$E(x_2 - \hat{x}_2) = \frac{x_2}{k} \tag{8.108}$$

Clearly the estimator for $\hat{x}_2$ is biased. However, the bias approaches
▶ zero as k tends to infinity.

PROBLEMS

1. Show that the estimators derived in Example 2, Section 8.2 (for any given autocorrelation function $\varphi(j - i)$) are unbiased.

2. Let the probability density of $y(i)$ be uniform over the interval $[0, x]$, i.e.,

$$p[y(i)] = \frac{1}{x} \qquad 0 \leq y(i) \leq x$$

Determine the maximum likelihood estimate of x given the independent samples $y(i)$; $i = 0, \ldots, k$. Is the estimator unbiased?

3. In the above problem replace the density function $p[y(i)] = \frac{1}{x}$ by

$$p[y(i)] = \frac{x^2 y(i) e^{-xy(i)}}{2}, \qquad y(i) > 0; \ x > 0$$

(a) Show that if $\hat{x}$ is the maximum likelihood estimate of x then

$$\hat{x} = \frac{2(k + 1)}{\sum_{0}^{k} y(i)}$$

(b) Determine whether this estimator is biased.

4. The autocorrelation of a scalar stationary random process $s(t)$ is given by

$$Es(t)s(t + \tau) = \varphi(\tau) = xe^{-|\tau|}$$

where x is unknown. Derive the least square estimate of x from the measurement $y(\tau)$, $0 \leq \tau \leq \tau_0$ where

$$y(\tau) = \int_0^T s(t)s(t + \tau) \, dt$$

5. Let the scalar random sequence $y(i)$ be an independent gaussian sequence with unknown mean x and unit variance. Derive a recursive estimator for the mean x.

6. Given the observation

$$y(i) = g'(i)x + v(i), \qquad i = 0, \ldots, k$$

where $v(i)$ is the observation noise. Derive the conditions the observation must satisfy in order that the least square estimate of x be unbiased.

7. Derive the estimator for x_1 and x_2 in Example 1, Section 8.4.

8. Discuss the similarity between the conditions necessary for complete observability discussed in Chapter 3 and that discussed in Section 8.5.

9. Derive a recursive procedure for "fitting" an observed signal $y(i)$, $i = 0, \ldots, k$ with the polynomial $x_0 + x_1e^{-i} + x_2e^{-2i}$. Show that the observability condition is always satisfied for this type of polynomial.

10. Let x, y and v be scalar functions satisfying

$$\dot{x}(t) = -x(t) + 0.1x^3(t)$$
$$y(\tau) = x(\tau) + v(\tau), \quad 0 \le \tau \le t$$

where $v(t)$ is white gaussian noise with unit spectral density. Derive the maximum likelihood estimate of x.

SELECTED READINGS

Albert, A. E. and L. A. Gardner, *Stochastic Approximation and Nonlinear Regression*, The MIT Press, Cambridge, Mass., 1967.

Bellman, R., H. Kagurada, R. Kalaba, and R. Sridhar, "Invariant Imbedding and Nonlinear Filtering," *J. Astronautical Sciences*, Vol. XIII, No. 3, May–June 1966.

Cramér, H., *Mathematical Methods of Statistics*, Princeton University Press, Princeton, N.J., 1946.

Helstrom, C. W., *Statistical Theory of Signal Detection*, Pergamon, New York, 1960.

Ho., Y. C., "On the Stochastic Approximation Method and Optimal Filtering," *J. Math. Analy. Appl.*, 6, 1962.

Middleton, D., *An Introduction to Statistical Communication Theory*, McGraw-Hill, New York, 1960.

chapter 9 / SYSTEM IDENTIFICATION

9.1 Introduction

The object of system identification is the determination of a mathematical model characterizing the operation of a system in some approximate form. The word system here implies a dynamic object where the variations of its states or outputs are governed by a set of physical laws and influenced by inputs to the object. The available information is typically the measured outputs or some function of the outputs, which in general may be contaminated by measurement noise. The inputs may be known functions applied for the purpose of identification, or unknown functions which it may be possible to monitor, approximately, or some combination of the two. Usually, before any measurement is made, some *a priori* information concerning the system is available. For example, it may be known whether the system is linear or nonlinear, discrete or continuous, etc. In other applications, the system may be known completely except for a finite set of parameters.

The areas of application of system identification are numerous, and a few are cited here. For example, it may be desired to determine a mathematical model for a human pilot in an appropriate environment, or to determine the parameters of a differential equation representing the dynamics of a ballistic missile entering through the atmosphere. In adaptive control systems, where the control is adapted to compensate for the effect of the environment on the system, it is clearly necessary to identify the state or parameters of the system during operation.

220

In this chapter, selected procedures for system identification will be introduced. These procedures are chosen on the basis of the generality of the approach and their feasibility in practical situations. Section 2 discusses a number of simple procedures which are not claimed to be optimal in any sense. However, the achievement of optimal system identification, at least in the maximum likelihood (or least-square curve-fitting sense), is our major concern to which the remainder of the chapter is devoted.

9.2 Linear Constant-Coefficient Systems

Let the *a priori* knowledge concerning the system be that it is a linear constant-coefficient system with zero initial condition and with scalar input and output given by $u(t)$ and $y(t)$, respectively. Such a system is completely characterized by the impulse response $h(t)$ or the transfer function $H(s)$ where

$$y(t) = \int_0^t h(t - \tau)u(\tau)\,d\tau \qquad (9.1)$$

If $u(t)$ and $y(t)$ are known, (9.1) can be solved for $h(t)$, although such a solution is seldom easy to carry out. If $u(t)$ is chosen to be the Dirac delta function, the observed output $y(t)$ becomes equal to $h(t)$, the desired function. Usually it is not possible to employ this method, since generation of a delta function as input is not feasible in practice. Moreover, any reasonable approximation to a delta function constitutes such a severe input to the system that it may cause the system to exhibit a behavior other than its usual form under normal conditions.* A procedure to eliminate this problem involves choosing an acceptable input of finite amplitude, such as e^{-t}, $t \geq 0$. Since this input has a Laplace transform $\dfrac{1}{s + 1}$, passing it through a system with transfer function $(s + 1)$ results in a delta function input to $H(s)$ as shown in Figure 9.1.

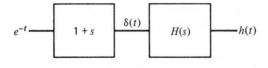

figure 9.1

---

* Most practical linear systems are actually the result of approximating a nonlinear system by a linear one. When driven by such severe inputs, these linear systems usually exhibit their nonlinear characteristics, so that analytical methods based on the assumption of linearity are no longer valid.

Now interchanging the operations of $1 + s$ and $H(s)$ yields the following

$$y(t) = \int_0^t h(t - \tau)e^{-\tau}\, d\tau = \int_0^t h(\delta)e^{\delta - t}\, d\delta \qquad (9.2)$$

$$h(t) = y(t) + \dot{y}(t) \qquad (9.3)$$

Equations (9.2) and (9.3) clearly yield $h(t)$, the impulse response, without having to use an impulse input. However, a practical consideration here is that $y(t)$ is usually contaminated by noise so that the differentiation in (9.3) may introduce significant error in measurement of $h(t)$.

Another procedure is to choose $u(t) = \sin \omega t$ and let the output $y(t)$ reach steady state,* which will usually be sinusoidal. The ratio of steady-state output amplitude to input amplitude may be obtained as a function of frequency, $\alpha(\omega)$. The steady-state phase shift $\theta(\omega)$ may also be obtained as a function of frequency. The transfer function $H(s)$ may then be represented, for steady-state sinusoidal inputs, as

$$H(j\omega) = \alpha(\omega)e^{j\theta(\omega)} \qquad (9.4)$$

By repeating this operation for a number of values of ω, a reasonable approximation to $H(j\omega)$ can be obtained. Knowing $H(j\omega)$ it is usually possible to obtain the impulse response $h(t)$ if it is needed.

When the input $u(t)$ is zero-mean white noise, with autocorrelation function $\delta(\tau)$, the following relationship can be mechanized to yield a value for $h(t_i)$, i.e., the system impulse response at $t = t_i$:

$$h(t_i) = Ey(t)u(t - t_i) \qquad (9.5)$$

In order to verify (9.5), let us substitute for $y(t)$ from (9.1):

$$h(t_i) = E\int_0^t h(t - \tau)u(\tau)u(t - t_i)\, d\tau \qquad (9.6)$$

Taking the expectation inside the integral sign yields the desired result. The operation given by (9.5) can be simultaneously performed for a number of values of t_i, thereby yielding a reasonable approximation to $h(t)$. In practice the expectation in (9.5) is approximated by

$$Ey(t)u(t - t_i) \approx \frac{1}{T}\int_0^T y(t)u(t - t_i)\, dt \qquad (9.7)$$

representing multiplication of $y(t)$ and $u(t)$ delayed by t_i and averaging over the interval $[0, T]$.

* That is, when the transients have died out. Clearly this limit exists only if the transients do indeed die out, meaning that the system is required to be stable.

9.3 Optimal Estimation of Impulse Response

The procedures for determination of the system impulse response described in the preceding section are merely measurement procedures without a claim of optimality in any sense. In this section, an approach to optimal estimation of a discrete approximation to the impulse response for a given input to the system will be introduced.

Let the impulse response $h(t)$, the input $u(t)$, and the output $g(t)$ be approximated by the sequences $h(t_i)$, $u(t_i)$, and $g(t_i)$, denoted by $h(i)$, $u(i)$, and $g(i)$, respectively. These quantities are related to each other by the convolution sum*

$$g(i + 1) = \sum_{j=0}^{i} h(j)u(i - j); \qquad g(0) = 0 \tag{9.8}$$

Clearly, the discrete approximation can be improved to any desired degree by simply sampling the functions $h(t)$, $u(t)$ and $g(t)$ more often. Let the observation be given by

$$y(i) = g(i) + v(i), \qquad 1 \le i \le I \tag{9.9}$$

where $v(i)$ is stationary observation noise. The purpose of using the discrete approximation is so that we only have to estimate a finite number of parameters, namely $h(0), \ldots, h(I - 1)$.

If the observation noise is gaussian with zero-mean, the results of Chapter 8 are directly applicable to the derivation of the maximum likelihood estimate of the parameters $h(i)$ given the observation (9.9). For example, if $v(i)$ is white gaussian noise, then the maximum likelihood estimate $\hat{h}(i)$ will satisfy the least square condition (8.13), which is written here as

$$\min_{h(j),\, j=0,\ldots,I-1} \sum_{i=0}^{I-1} \left[y(i - 1) - \sum_{j=0}^{i} h(j)u(i - j) \right]^2 \tag{9.10}$$

Let us introduce the vectors y_1 and h_1 and the matrix U_1 defined as follows

$$h_I = \begin{bmatrix} h(0) \\ h(1) \\ \cdot \\ \cdot \\ \cdot \\ h(I - 1) \end{bmatrix} \tag{9.11}$$

---

* In (9.8) the initial condition of the system, and consequently the response to the initial condition, is assumed zero.

$$y_I = \begin{bmatrix} y(1) \\ \cdot \\ \cdot \\ \cdot \\ y(I) \end{bmatrix} \qquad (9.12)$$

$$v_I = \begin{bmatrix} v(1) \\ \cdot \\ \cdot \\ \cdot \\ v(I) \end{bmatrix} \qquad (9.13)$$

$$U_I = \begin{bmatrix} u(0) & 0 & \cdots & & & \\ u(1) & u(0) & 0 & \cdots & & \\ u(2) & u(1) & u(0) & 0 & \cdots & \\ \cdot & & & & & \\ \cdots\cdots\cdots\cdots\cdots\cdots\cdots\cdots\cdots & & & & & \\ \cdot & & & & & \\ u(I-1) & u(I-2) & \cdots & \cdots & u(0) \end{bmatrix} \qquad (9.14)$$

Since the matrix U_I is a triangular matrix, it will certainly be nonsingular as long as $u(0)$ is not zero. Using the above notation, (9.9) can be written as

$$y_I = U_I h_I + U_I \qquad (9.15)$$

Hence (9.10) becomes

$$\min_{h_I} [y_I - U_I h_I]'[y_I - U_I h_I] \qquad (9.16)$$

carrying out the minimization by means of direct differentiation results in

$$\hat{h}_I = U_I^{-1} y_I \qquad (9.17)$$

The result (9.16) was expected, since (9.9) yields I equations in terms of the I unknowns $h(0), \ldots, h(I-1)$. The covariance matrix of the estimation error $h_I - \hat{h}_I$ can now be obtained.

$$\text{cov}\,(h_I - \hat{h}_I) \triangleq E(h_I - \hat{h}_I)(h_I - \hat{h}_I)' \qquad (9.18)$$

Substituting for h_I and $\hat{h}_I$ from (9.15) and (9.16) yields

$$E(h_I - \hat{h}_I)(h_I - \hat{h}_I)' = U_I^{-1} K U_I'^{-1} \qquad (9.19)$$

where K is a diagonal matrix representing the covariance function of the observation noise vector v_I.

The above estimation procedure assumes that we have no *a priori* information concerning the impulse response $h(i)$. It is possible that the values of $\hat{h}(i)$ may still be significant and may vary considerably as a function of i even for large values of i. On the other hand, in many applications we may know that

$$h(i) \approx 0 \qquad i > J$$

even though we may wish to take observations for $i = 1, \ldots, I; I > J + 1$. In order to make use of this additional information, at least in an approximate sense, we may impose the condition

$$h(i) = 0 \qquad i > J$$

on $h(i)$. Let us define the matrix U_J and vector h_J.

$$U_J = \begin{bmatrix} u(0) & 0 & \cdots & \\ & & & \\ u(1) & u(0) & \cdots & \\ \vdots & & & \\ u(J) & \cdots & & u(0) \\ \vdots & \cdots & & \\ \vdots & \cdots & & \\ \vdots & \cdots & & \\ u(I-1) & \cdots & & u(I-J-1) \end{bmatrix} \qquad (9.20)$$

$$h_J = \begin{bmatrix} h(0) \\ \vdots \\ h(J) \end{bmatrix} \qquad (9.21)$$

Notice that the matrix U_J is of the order $(I + 1) \times (J + 1)$ and has rank $J + 1$ if $u(0)$ is not zero. From (9.8) and (9.9), we then have

$$y_I = U_J h_J + v_I \qquad (9.22)$$

and (9.10) becomes

$$\min_{h_J} [y_I - U_J h_J]'[y_I - U_J h_J] \qquad (9.23)$$

Carrying out the minimization by differentiation yields

$$\hat{h}_J = [U_J' U_J]^{-1} U_J' y_I \qquad (9.24)$$

The $(J + 1) \times (J + 1)$ matrix $U_J'U_J$ is nonsingular, since U_J has a rank equal to $J + 1$. The covariance function of $(h_J - \hat{h}_J)$ is obtained by using (9.22) and (9.24)

$$E[h_J - \hat{h}_J][h_J - \hat{h}_J]' = [U_J'U_J]^{-1}U_J'KU_J[U_J'U_J]^{-1} \quad (9.25)$$

where K is the covariance matrix of the observation noise vector U_J. If K is the identity matrix, (9.25) reduces to

$$E(h_J - \hat{h}_J)(h_J - \hat{h}_J)' = [U_J'U_J]^{-1}$$

Since we have concentrated on estimating only a part of the parameters $h(0), \ldots, h(I - 1)$, namely $h(0), \ldots, h(J)$, it is expected that an improved covariance function of $h_J - \hat{h}_J$ results. Furthermore, it is observed that each term of the covariance function given by (9.19) and (9.25) is reduced as the system input $u(i)$ is increased in magnitude (say by a scalar multiplier). This is expected because as the input $u(i)$, and hence the output $g(i)$, is increased in magnitude, it tends to overcome the additive observation noise. Moreover, this fact suggests that an optimal input $u(i)$, in the light of achieving a better estimate of impulse response, may exist such that some function of the error covariance matrix (e.g., the trace of the matrix) is minimized. The following example illustrates these points.

▶ **example 1**
A linear time-invariant system is driven by a step function (i.e., $u(i) = 1$; $i \geq 0$). The output $g(i)$ is observed for $i = 1, 2, 3$ and is contaminated by additive white noise $v(i)$ where

$$y(i) = g(i) + v(i)$$
$$Ev(i) = 0$$
$$Ev(i)v(j) = \Delta(j - i)$$

It is desired (a) to determine the maximum likelihood (or least square) estimate of $h(0), h(1), h(2)$, and (b) to determine estimates of $h(0)$ and $h(1)$, without estimating $h(2)$.

The estimation procedure is straightforward [using (9.17)] and is left as an exercise. Let us concentrate on the covariance function

$$E(h_1 - \hat{h}_1)(h_1 - \hat{h}_1)'$$

where

$$h_J = h_1 = \begin{bmatrix} h(0) \\ h(1) \end{bmatrix}; \quad h_I = h_2 = \begin{bmatrix} h(0) \\ h(1) \\ h(2) \end{bmatrix}$$

is to be evaluated for parts (a) and (b). Since $K = I$ in (9.19) and

$$U_I = \begin{bmatrix} 1 & 0 & 0 \\ 1 & 1 & 0 \\ 1 & 1 & 1 \end{bmatrix}$$

we have

$$E(h_I - \hat{h}_I)(h_I - \hat{h}_I)' = \begin{bmatrix} 1 & -1 & 0 \\ -1 & 2 & -2 \\ 0 & -1 & 1 \end{bmatrix}$$

hence

$$E(h_J - \hat{h}_J)(h_J - \hat{h}_J)' = \begin{bmatrix} 1 & -1 \\ -1 & 2 \end{bmatrix} \tag{9.26}$$

This is the covariance function obtained in part (a) where we are estimating $h(0)$, $h(1)$, and $h(2)$. On the other hand, from (9.25), and since

$$U_J = \begin{bmatrix} 1 & 0 \\ 1 & 1 \\ 1 & 1 \end{bmatrix}$$

we have

$$E(h_J - \hat{h}_J)(h_J - \hat{h}_J)' = \begin{bmatrix} 1 & -1 \\ -1 & 1.5 \end{bmatrix} \tag{9.27}$$

Comparison of (9.26) and (9.27) reveals that the variance $Eh^2(1)$ obtained in part (b) is smaller than the corresponding value in part (a) (1.5 ▶ vs. 2).

Let us now discuss the influence of the system input on the estimation of h_J. The input $u(i)$ was chosen arbitrarily as a constant for this problem. Suppose we are to design an optimal input sequence $u * (i)$ minimizing some meaningful function (such as the trace) of the error covariance matrix. To avoid a trivial solution, let us impose an energy constraint given by $\sum_{i=1}^{2} u^2(i) \leq 1$. The covariance function given by (9.25) for part (b) of the example in terms of the unknown inputs $u(0)$, $u(1)$, $u(2)$ is

$$E[h_J - \hat{h}_J][h_J - \hat{h}_J]' = \begin{bmatrix} \sum_{0}^{2} u^2(i) & u(1)u(0) + u(2) \\ u(1)u(0) + u(2) & \sum_{0}^{1} u^2(i) \end{bmatrix}^{-1}$$

$$= \frac{1}{\Delta} \begin{bmatrix} \sum_{0}^{1} u^2(i) & -u(1)u(0) + u(2) \\ -u(1)u(0) + u(2) & \sum_{0}^{2} u^2(i) \end{bmatrix}$$

where

$$\Delta = u^4(0) + u^2(0)u^2(2) + u^4(1) - 2u^2(1)u(0)u(2) + u^2(0)u^2(1)$$

Suppose we decide to minimize the trace of this matrix, that is

$$\min_{u(0),u(1),u(2)} E[(h(0) - \hat{h}(0)) + (h(1) - \hat{h}(1))]^2$$

$$\sum_{i=0}^{2} u^2(i) \leq 1$$

Using a Lagrange multiplier λ,* we replace this by

$$\min_{u(0),u(1),u(2)} E[(h(0) - h(0))^2 + (h(1) - h(1))]^2 + \lambda[u^2(0) + u^2(1) + u^2(2) - 1]$$

Differentiating with respect to $u(0)$, $u(1)$, and $u(2)$ yield a set of three simultaneous equations which are to be solved for values of $u(i)$ in terms of λ. Finally, λ is obtained by satisfying the energy constraint. This is a routine but tedious problem. If we had arbitrarily decided to set $u(2) = 0$, the solution would simplify considerably resulting in

$$u(0) = \tfrac{1}{2}$$
$$u(1) = \tfrac{1}{2}$$

9.4 Parameter Estimation, Linear Systems

The procedures of the preceding sections require very little *a priori* knowledge of the system impulse response. In a majority of practical applications, additional information, such as the order of the system or the form of impulse response or system transfer function, are available. This simply means that the impulse response function is known within a finite set of parameters. The problem of estimation of impulse response will then logically be replaced by the problem of estimating the unknown parameters.

Let the impulse response be given by $h(t, x)$ where x is a vector representing the unknown parameters. The quantity $g(t, x)$ is defined by

$$g(t, x) = \int_0^t h(t - \tau, x)u(\tau) \, d\tau \qquad (9.28)$$

The observation is given by

$$y(\tau) = g(\tau, x) + v(\tau) \qquad 0 \leq \tau \leq t \qquad (9.29)$$

where $v(t)$ is the additive observation noise. The most likely value of x given the observations $y(\tau)$, $0 \leq \tau \leq t$, can be obtained by using the results of

* Sokolnikoff, I. S. and Redheffer, R. M., *Mathematics of Physics and Modern Engineering*, McGraw-Hill, 1958.

Section 8.2. If the observation noise is white gaussian, then the results of Section 8.3 are applicable.

▶ **example 1**
Let the impulse response of a system be given by

$$h(t) = e^{-xt}$$

and the observation by

$$y(t) = \int_0^t e^{-x(t-\tau)}u(\tau)\,d\tau + v(t) \qquad 0 \le \tau \le t \qquad (9.30)$$

where $v(t)$ is white gaussian with zero-mean and covariance function given by

$$Ev(t_2)v(t_1) = \delta(t_2 - t_1)$$

Choosing a value for $u(t)$, such as unity, (9.30) yields

$$y(\tau) = \frac{1 - e^{-x\tau}}{x} + v(\tau) \qquad 0 \le \tau \le t \qquad (9.31)$$

Using the results of Section 8.3, since $v(t)$ is white noise, the maximum likelihood estimate of x will satisfy the least square condition of (8.14)

$$\min_x \int_0^T \left[y(t) - \frac{1 - e^{-xt}}{x} \right]^2 dt \qquad (9.32)$$

Denoting the estimate by $\hat{x}$ and carrying out the differentiation with respect to x, it follows that

$$\int_0^T \left[y(t) - \frac{1 - e^{-\hat{x}t}}{\hat{x}} \right]\left[\frac{e^{-\hat{x}t}(1 + t\hat{x}) - 1}{\hat{x}^2} \right]\alpha t = 0 \qquad (9.33)$$

This is the likelihood equation. Unfortunately, it is in a form which by no means is easy to solve for the optimal estimate $\hat{x}$, even in the case of the present simple system. As it turns out, the minimization indicated in (9.32) is easier to carry out, numerically, in comparison with the steps needed to ▶ solve the likelihood equation (9.33).

The preceding example is discussed in a more general setting in Section 9.5. It is clear that the main difficulty in solving (9.33) is in the fact that the impulse response e^{-xt} is a nonlinear (in this case, transcendental) function of the time constant $\left(\frac{1}{x}\right)$. The case where this relationship appears in linear form clearly avoids the above difficulty. The following example is an illustration of this point.

▶ **example 2**

It is desired to identify, in the maximum likelihood sense, the system given by the transfer function

$$H(s) = \frac{K(s + a)}{(s + 1)(s + 2)}$$

where the parameters K and a are unknown. The input is chosen arbitrarily as unit step, and the observation noise is given by $v(t)$ defined in Example 1. The impulse response of the system can be written as

$$h(t) = x_1 e^{-t} + x_2 e^{-2t}$$

where

$$x_1 = K(a - 1)$$
$$x_2 = K(2 - a)$$

We decide to obtain maximum likelihood estimates of x_1 and x_2 in place of K and a. Using the results of Section 8.3, the maximum likelihood estimates $\hat{x}_1$ and $\hat{x}_2$ accomplish the following minimization

$$\min_{x_1, x_2} \int_0^t \left[y(\tau) - x_1(1 - e^{-\tau}) - \frac{x_2}{2}(1 - e^{-2\tau}) \right]^2 d\tau$$

Differentiating with respect to x_1 and x_2 and equating the derivatives to zero yields the likelihood equations

$$\hat{x}_1 \int_0^t (1 - e^{-\tau})^2 \, d\tau + \hat{x}_2 \int_0^t \tfrac{1}{2}(1 - e^{-2\tau})(1 - e^{-\tau}) \, d\tau = \int_0^t y(\tau)(1 - e^{-\tau}) \, d\tau$$

$$\hat{x}_1 \int_0^t (1 - e^{-\tau})(1 - e^{-2\tau}) \, d\tau + \hat{x}_2 \int_0^t \tfrac{1}{2}(1 - e^{-2\tau}) \, d\tau = \int_0^t y(\tau)(1 - e^{-2\tau}) \, d\tau$$

The integrations on the left hand side do not involve the data, $y(t)$, so they may be performed at once to obtain

$$[-\tfrac{3}{2} + t + 2e^{-t} - \tfrac{1}{2}e^{-2t}]\hat{x}_1 + \left[-\tfrac{7}{12} + \frac{t}{2} + \tfrac{1}{2}e^{-t} - \tfrac{1}{4}e^{-2t} - \tfrac{1}{6}e^{-3t} \right]\hat{x}_2$$
$$= \int_0^t y(\tau)(1 - e^{-\tau}) \, d\tau$$

$$[-\tfrac{7}{6} + t + e^{-t} + \tfrac{1}{2}e^{-2t} - \tfrac{1}{3}e^{-3t}]\hat{x}_1 + \left[-\tfrac{3}{8} + \frac{t}{2} + \tfrac{1}{2}e^{-2t} - \tfrac{1}{8}e^{-4t} \right]\hat{x}_2$$
$$= \int_0^t y(\tau)(1 - e^{-2\tau}) \, d\tau$$

Finally, these simultaneous equations can easily be solved to yield $\hat{x}_1$ and $\hat{x}_2$. As an alternative method, we may use results of Section 8.4 to ▶ obtain a recursive estimator for x_1 and x_2.

9.5   Parameter Estimation, Nonlinear Systems

Identification of a general nonlinear system without *any a priori* knowledge
concerning the equations governing the system (similar to the procedures of
Sections 9.2 and 9.3) is clearly not feasible. Even if the order of the differential
equation is known, the many forms which a nonlinear system can assume
makes any effort toward system identification fruitless. In practice, all one
can expect is to derive techniques for estimating a finite number of parameters
in an otherwise completely known system.

The problem is formulated as follows. The system equations are given by

$$\dot{x}(t) = f[x(t), t, \gamma] \tag{9.34}$$

where x is an n-vector and γ is an r-vector representing the unknown parameters. The observation $y(\tau)$, $0 \le \tau \le t$, is given by

$$y(t) = g[x(t), t] + v(t) \tag{9.35}$$

One approach to the problem is to augment, as discussed in Section 2.5, the
set of equations given by (9.34) in order to absorb the unknown parameters γ
as additional states in the x-vector. If the *a priori* probability density function
of γ, together with the initial condition $x(0)$, is known, the system and
observations can be approximated by discrete equations, the procedures of
Chapter 7 become directly applicable, and one may obtain the posterior mode
estimate of γ. When no *a priori* statistics concerning the parameter γ are
available, the least square estimate $\hat{\gamma}$ can be chosen which satisfies (for scalar
observation)

$$\min_{\gamma} J(\gamma) = \min_{\gamma} \int_0^t [y(\tau) - g(x(\tau), t)]^2 \, d\tau \tag{9.36}$$

As shown in Chapter 8, this estimate becomes equivalent to the maximum
likelihood estimate of γ when the additive observation noise is white gaussian.

The minimization indicated in (9.36), together with the differential equation
(9.34), represents a purely deterministic operation. A recursive solution to
this problem was obtained in Section 8.8.*

A direct approach in carrying out the minimization in (9.36) is also
possible and is referred to as "model matching." The idea simply is to
simulate equation (9.34) (say with a given initial condition) on a computer
and evaluate (9.36) for a first guess at the value of γ denoted by γ°. This
results in a value for (9.36) denoted by $J(\gamma^\circ)$. The successive values of γ^i,
$i \ge 0$ are chosen such that

$$J(\gamma^{i+1}) < J(\gamma^i) \tag{9.37}$$

* In order to put the problem in the form discussed in Section 8.8 the system equation (9.34)
should be augmented to include the parameter γ. *See* Section 3.5.

In the case of a scalar parameter, the successive values of γ^i can simply be chosen according to (9.37) until a local minimum for $J(\gamma)$ is obtained. Whether this also is a global minimum, at least to the best of our knowledge, depends on the convexity* of the function $J(\gamma)$. In practice, when this information is not available, enough different initial choices for γ^0 can be chosen until the operator is reasonably certain the absolute minimum of $J(\gamma)$ has been reached.

Although the above procedure is applicable to the vector-valued γ, it is computationally very inefficient (because of the large number of different component values of γ which must be tried) and we should seek a more organized way to choose the successive values of γ^i. One approach is to apply the above procedure to each component of γ at a time and then iterate over all components until the minimum of $J(\gamma)$ is achieved. This is called the relaxation method.

A more appealing procedure is the method of steepest descent. Let us choose γ^{i+1} in the vicinity of γ^i such that $J(\gamma^{i+1})$ is as small as possible. This is similar to the problem we started to solve but over small ranges of values of γ. Consequently we can linearize $J(\gamma)$ in the neighborhood of γ^i and perform the minimization. This results in the steepest descent procedure and is implemented by setting

$$\gamma^{i+1} = -\alpha \frac{\partial J(\gamma)}{\partial \gamma}\bigg|_{\gamma=\gamma^i} + \gamma^i \tag{9.38}$$

where the term $\dfrac{\partial J(\gamma)}{\partial \gamma}$ is the gradient† of the function $J(\gamma)$. Obviously, $J(\gamma)$ should be differentiable in γ in order for the gradient to exist. The term α is a scalar multiplier chosen as small as necessary to make the successive values of $J(\gamma^i)$ monotonically decreasing. Clearly, the convergence will be impeded if α is chosen too small. The best choice for α is usually obtained by a trial and error procedure. Again, the convergence to the global minimum of the function $J(\gamma)$ is assured if $J(\gamma)$ is convex.

If the minimization of $J(\gamma)$ is to be carried out on an analog computer, (9.38) may be substituted by the following differential equation

$$\dot{\gamma} = -\alpha \frac{\partial J(\gamma)}{\partial \gamma} \tag{9.39}$$

Here we attempt to generate a continuous function of time $\gamma(t)$ which tends to γ.

* The condition of convexity assures that $J(\gamma)$ has only one minimum and hence the local minimum is also a global minimum. This condition is stated as $\alpha J(\gamma_1) + (1-\alpha)J(\gamma_2) > J[\alpha\gamma_1 + (1-\alpha)\gamma_2]$, for all γ_1, γ_2, and $1 > \alpha > 0$.
† *See* Section 1, Chapter 1.

The major difficulty in application of the steepest descent method is derivation of the gradient vector $\dfrac{\partial J(\gamma)}{\partial \gamma}$. From (9.36) we have

$$\frac{\partial J(\gamma)}{\partial \gamma} = -2 \int_0^t [y(\tau) - g(x(\tau), \tau)] \left[\frac{\partial x(\tau)}{\partial \gamma}\right]' \left[\frac{\partial g(x(\tau), \tau)}{\partial x(\tau)}\right] d\tau \qquad (9.40)$$

All of the terms under the integral sign are known except the $r \times n$ matrix $\dfrac{\partial x(\tau)}{\partial \tau}$. Let us differentiate both sides of (9.34) with respect to γ.

$$\frac{\partial \dot{x}(t)}{\partial \gamma} = \frac{\partial}{\partial \gamma} f[x(t), t, \gamma] \qquad (9.41)$$

Changing the order of differentiation on the left hand side of (9.41), and carrying out the differentiations, we obtain

$$\frac{d}{dt}\left[\frac{\partial x(t)}{\partial \gamma}\right] = \frac{\partial f[x(t), t, \gamma]}{\partial \gamma} + \left[\frac{\partial f[x(t), t, \gamma]}{\partial x(t)}\right]' \frac{\partial x(t)}{\partial \gamma} \qquad (9.42)$$

Equation (9.42) is a linear differential equation in terms of $\dfrac{\partial x(t)}{\partial \gamma}$ and has the forcing function $\dfrac{\partial f[x(t), t, \gamma]}{\partial \gamma}$. Since the initial condition $x(0)$ of (9.34) is independent of γ the initial condition for (9.42) becomes

$$\left.\frac{\partial x(t)}{\partial \gamma}\right|_{t=0} = 0 \qquad (9.43)$$

Finally, (9.40) and (9.42) can be simulated to generate the gradient vector $\dfrac{\partial J(\gamma)}{\partial \gamma}$ which in turn can be used in equations (9.38) or (9.39) describing the steepest descent procedure. Equation (9.42) is referred to as the "sensitivity equation."

▶ example 1

The unit step response of a linear constant-coefficient scalar system is described by the solution of the following differential equation,

$$\dot{x} = -\gamma x + \gamma \qquad (9.44)$$

where $1/\gamma$ is an unknown time constant. It is desired to derive the least square estimate of γ by using the observation

$$y(\tau) = x(\tau) + v(\tau), \qquad 0 \le \tau \le t$$

The sensitivity equation becomes

$$\frac{d}{dt}\left[\frac{\partial x(t)}{\partial \gamma}\right] = \gamma^i\left[\frac{\partial x(t)}{\partial \gamma}\right] + 1 + x(t) \qquad (9.45)$$

$$\frac{\partial x(0)}{\partial \gamma} = 0$$

where γ^i is the ith guess at the least square estimate $\hat{\gamma}$. Equation (9.40) becomes

$$\frac{\partial J(\gamma)}{\partial \gamma}\bigg|_{\gamma=\gamma^i} = -2\int_0^t [y(\tau) - x(\tau)]\frac{\partial x(\tau)}{\partial \gamma}\,d\tau \qquad (9.46)$$

Starting with ith approximation to $\hat{\gamma}$, namely γ^i, (9.44) is solved for $x(t)$. The sensitivity term $\dfrac{\partial x(t)}{\partial \gamma}$ is obtained by using this to obtain the forcing function in (9.45). Using $\dfrac{\partial x(t)}{\partial \gamma}$ and $x(t)$ in (9.46) the gradient $\dfrac{\partial J(\gamma)}{\partial \gamma}\bigg|_{\gamma=\gamma^i}$ is obtained. Finally, (9.38) is used to derive the next guess at $\hat{\gamma}$, namely γ^{i+1}. It is interesting to note that equations (9.44), (9.45), and (9.46) can
▶ run simultaneously.

▶ **example 2**
We wish to derive the least square estimate of the parameter γ in the system governed by

$$\dot{x} = -\gamma \sin x + 0.1$$
$$x(0) = 0 \qquad (9.47)$$

The observation is given by

$$y(\tau) = x(\tau) + v(\tau) \qquad 0 \le \tau \le t$$

The sensitivity equation (9.42) becomes

$$\frac{d}{dt}\left[\frac{\partial x(t)}{\partial \gamma}\right] = -[\gamma^i \cos x(t)]\frac{\partial x(t)}{\partial \gamma} - \sin x(t) \qquad (9.48)$$

where $x(t)$ is obtained from (9.47) for $\gamma = \gamma^i$. The solution $\dfrac{\partial x(t)}{\partial \gamma}$ of (9.48) along with $x(t)$ is substituted into the gradient equation (9.40)

$$\frac{\partial J(\gamma)}{\partial \gamma} = -2\int_0^t [t(\tau) - x(\tau)]\frac{\partial x(\tau)}{\partial \gamma}\,d\tau$$

Finally, (9.38) is used to derive γ^{i+1}. The iteration is continued until no
▶ further improvement in $J(\gamma)$ is obtained.

PROBLEMS

1. Derive the least square estimators for (a) $h(0)$, $h(1)$, $h(2)$, and (b) $h(0)$, $h(1)$ in Example 1, Section 9.3.

2. Let the impulse response of a constant-coefficient system by given by

$$h(t) = e^{-xt}$$

The input to the system is given by $u(t) = \delta(t)$. The output contains additive white gaussian noise. Derive the maximum likelihood estimate of x.

3. Discuss the solution in Problem 2 above if $u(t) = e^{-t}$. Obtain an approximate estimate of x truncating the Taylor series expansion of the exponential term and keeping the first 3 terms.

4. Derive an equation yielding the least square estimate of the scalar parameter γ in the nonlinear differential equation

$$\dot{x} = \gamma x - x^3 + 1$$
$$x(0) = 0$$

The observation y is given by

$$y(\tau) = x(\tau) + v(\tau), \qquad 0 \leq \tau \leq t$$

where $v(\tau)$ is the observation noise.

SELECTED READINGS

Eveleigh, V. W., *Adaptive Control and Optimization Techniques*, McGraw-Hill, New York, 1967.

Kalman, R. E., "Design of a Self-Optimizing Control System," Transactions ASME, *J. Basic Engineering*, **80**, February, 1958.

Levin, M. J., "Optimum Estimation of Impulse Response in the Presence of Noise," *IRE Transactions on Circuit Theory*, March, 1960.

Tomovic, R., *Sensitivity Analysis of Dynamic Systems*, McGraw-Hill, New York, 1963.

Turin, G. L., "On the Estimation in the Presence of Noise of the Impulse Response of a Random, Linear Filter," *IRE Transactions on Information Theory*, March, 1957.

chapter 10 / SIGNALS OF KNOWN FORM IN THE PRESENCE OF NOISE

10.1 Introduction

There are many applications where the observation contains a signal with partially or completely known form. For example, in radar and various types of radio transmission, the useful signal at the receiver has a fixed waveform. When the signal waveform is known except for a finite number of parameters, this information should definitely be used by the estimator. In other words, we should concentrate on estimation of the unknown parameters, such as the amplitude or phase of an incoming sine wave corrupted by noise, if we desire to estimate the function itself. The procedures developed in the past chapters can be used for this purpose. On the other hand, when the signal waveform is completely known, its estimation is meaningless. However, we may be interested in obtaining knowledge of whether the signal is present in, or absent from, the observation. It is then of no concern to us whether the estimator produces a reasonable replica of the true signal, and consequently there is no justification for minimization of a cost function based on the estimation error or any of the other criteria discussed in the past chapters. Instead, we have to seek a measure related to the accuracy of the decision involving the presence or absence of a signal.

236

10.2 Signals with Partially Known Form

When the signal is known except for a finite number of parameters, and if no *a priori* statistics concerning these parameters are available, we can make use of the results of Chapter 8 to derive the maximum likelihood estimate of the parameters, as was done in the examples in Section 8.3. When the *a priori* statistics are available, an attempt should be made to represent the signal by a linear or nonlinear differential equation model so that any of the results of Chapters 4–7 become applicable. For example, when the signal is a ramp function given by $s(t) = \alpha_0 + \alpha_1 t$, it can be equivalently represented as

$$\dot{x}_1(t) = x_2(t)$$
$$\dot{x}_2(t) = 0 \qquad (10.1)$$
$$s(t) = x_1(t)$$

with the initial conditions $x_1(0) = \alpha_0$, $x_2(0) = \alpha_1$. If we now receive $y(t) = s(t) + v(t)$ where $v(t)$ is the observation noise, the estimation of the parameters α_0 and α_1 can be handled by the techniques of Chapter 4 or 5. As another example, suppose it is desired to obtain a least square estimate of the amplitudes a and b in an observed signal

$$y(t) = a \sin t + b \cos t + v(t) \qquad (10.2)$$

The observation $y(t)$ and the parameters a and b are equivalently represented by the following model

$$\dot{x}_1(t) = x_2(t)$$
$$\dot{x}_2(t) = -x_1(t)$$
$$y(t) = x_1(t) + v(t) \qquad (10.3)$$
$$x_1(0) = b$$
$$x_2(0) = a$$

It should be noted that, in general, it is not possible to represent any arbitrary signal as a solution of a differential equation (linear or nonlinear), because the arbitrary signal may fail to have the continuity and differentiability required of all such solutions. Furthermore, in many cases, such a representation does not necessarily simplify the estimation problem at hand.

10.3 Signals with Completely Known Form

Let $s(i)$ be a completely known signal corrupted by noise $v(i)$ yielding the observation

$$y(i) = s(i) + v(i) \qquad (10.4)$$

in the case the signal is present and

$$y(i) = v(i) \qquad (10.5)$$

in the absence of the signal. Having observed $y(i)$ over interval $0 \le i \le n$ we would like to determine whether the signal is present in or absent from the observation. In the absence of the signal, the observation $y(i)$ has a joint probability density function denoted here by

$$p[y(0), \ldots, y(n) \mid \text{signal absent}] \triangleq p_v[y(0), \ldots, y(n)] \qquad (10.6)$$

Note that (10.6) is conditioned on the absence of signal, and thus constitutes a likelihood function. Since the signal $s(i)$ is completely known, the joint probability density function of $y(i)$ in the presence of signal is also known and is denoted by

$$p[y(0), \ldots, y(n) \mid \text{signal present}] \triangleq p_s[y(0), \ldots, y(n)]$$
$$= p_v[y(0) - s(0), \ldots, y(n) - s(n)] \qquad (10.7)$$

Note that (10.7) is conditioned on the presence of signal, and thus constitutes another likelihood function. Because the noise $v(i)$ is always present, it will seldom be possible to decide with certainty whether the signal $s(i)$ is present or absent. Therefore, a reasonable choice is to decide the presence or absence of the signal according to whether, for a given observation sequence $y(0), \ldots, y(n)$, (10.7) or (10.6) is the larger. Equivalently, we may decide that the signal is present if the ratio $\dfrac{p_s}{p_v}$ is larger than a given value (called the "threshold"). This ratio is referred to as the likelihood ratio, since it is a ratio of two likelihood functions (10.7) and (10.6). Clearly, if the value to which the likelihood ratio is compared is large we would, on the average, become very conservative about deciding the presence of the signal and would be more willing to decide on its absence. This leads to a low probability of "false alarm." However, with a high threshold value, there is also a correspondingly high probability of deciding that the signal is absent when it is indeed present.

If the observation noise $v(i)$ is a zero-mean independent white gaussian sequence with covariance function given by

$$Ev(i)v(j) = l\Delta(j - i) \qquad (10.8)$$

The joint density functions given by (10.6) and (10.7) become

$$p_v[y(0), \ldots, y(n)] = \frac{1}{(2\pi)^{n/2}l^{1/2}} \exp - \frac{1}{2k} \sum_{i=0}^{n} y^2(i) \qquad (10.9)$$

and

$$p_s[y(0), \ldots, y(n)] = \frac{1}{(2\pi)^{n/2}l^{1/2}} \exp - \frac{1}{2k} \sum_{i=0}^{n} [y(i) - s(i)]^2 \qquad (10.10)$$

The likelihood ratio is then given by

$$\frac{p_s}{p_v} = \exp -\frac{1}{2l}\sum_{i=0}^{n}[s^2(i) - 2s(i)y(i)] \tag{10.11}$$

Consequently, the right-hand side of (10.11) is to be compared with some given threshold value, say k, in order to decide whether or not the signal is present. Since the $s(i)$ are assumed to be known numbers, it is customary to make some simplifications in the actual threshold-comparison process. Since the exponential function in (10.11) is monotonic in its argument, we can perform an equivalent threshold operation if we compare the argument $-\frac{1}{2l}\sum_{i=0}^{n}$ $[s^2(i) - 2s(i)y(i)]$ with $\log k$. Similarly, we can compare $\sum_{i=0}^{n}[s^2(i) - 2s(i)y(i)]$ with $-2l\log k$. Using the fact that $\sum_{i=0}^{n}s^2(i)$ is known, we can proceed to obtain, finally, the simplest version of the threshold comparison process:

$$\sum_{i=0}^{n}s(i)y(i) \quad \begin{matrix} \geq c, \text{ decide signal is present} \\ < c, \text{ decide signal is absent} \end{matrix} \tag{10.12}$$

where c is the modified threshold given by

$$c \triangleq l\log k + \frac{1}{2}\sum_{i=0}^{n}s^2(i)$$

If we identify i with time sample t_i such that $t_0 = 0$ and $\lim_{i\to\infty} t_i = T$, then (10.12) yields in the limit

$$\int_0^T s(t)y(t)\,dt \quad \begin{matrix} \geq c \text{ decide signal is present} \\ < c, \text{ decide signal is absent} \end{matrix} \tag{10.13}$$

which is the continuous version of the threshold comparison process. The physical interpretation of (10.12) and (10.13) is as follows: The decision on whether the signal is present or absent can be made by a comparison of the cross correlation of the observation $y(t)$ and the known signal $s(t)$, over the interval of observation, with a known constant c (the threshold), where c does not depend on the observations $y(t)$. In Section 10.5 it will be shown that a linear system can be derived to yield the desired operation given by (10.13).

10.4 Maximization of (Expected) Signal-to-Noise Ratio

Another reasonable measure on which the decision concerning the presence or absence of a known signal $s(t)$ in the observation $y(t)$ can be based is the signal-to-noise power ratio at the time (t_0) when the decision is to be made.

The use of such a power ratio is easy to justify since if no signal is present the signal-to-noise power ratio is zero, and if the signal to noise is very large it is very likely that the signal is present. The signal power is defined as

$$S \triangleq s^2(t_0) \tag{10.14}$$

which is the square of a deterministic quantity $s(t_0)$, and the noise power is defined as

$$N \triangleq Ev^2(t) \tag{10.15}$$

which is the expected value of the square of a random quantity $v(t)$. In general, the signal-to-noise ratio $\rho \triangleq S/N$ obtained from (10.14) and (10.15) may be very small. As a matter of fact, it will be zero if $v(t)$ is white noise since $Ev^2(t) = \infty$ and $s^2(t_0)$ is finite. However, there is the possibility of improving the signal-to-noise ratio if we pass the observation through a filter. If we arbitrarily decide this filter to be linear constant-coefficient, it can then be described by its Fourier transfer function $H(j\omega)$. If $Y(j\omega)$ and $X(j\omega)$ are Fourier transforms of the input and output of the filter respectively, we have

$$X(j\omega) = H(j\omega)Y(j\omega) \tag{10.16}$$

When both signal $s(t)$ and noise $v(t)$ are present, and assuming that $s(t)$ and $v(t)$ are additive, the output $x(t)$ can be written as

$$x(t) = x_s(t) + x_v(t) \tag{10.17}$$

where $x_s(t)$ and $x_v(t)$ are the responses to $s(t)$ and $v(t)$ respectively. Therefore, the signal-to-noise ratio at the output at time t_0 is given by

$$\rho_0 \triangleq \frac{S_0}{N_0} = \frac{x_s^2(t_0)}{Ex_v^2(t)} \tag{10.18}$$

Let us now obtain the linear filter $H(j\omega)$ which maximizes (10.18). From (10.16) we have

$$X_s(j\omega) = H(j\omega)S(j\omega) \tag{10.19}$$

where $S(j\omega)$ is the Fourier transform of $s(t)$. Hence,

$$x_s(t) = \frac{1}{2\pi} \int_{-\infty}^{+\infty} H(j\omega)S(j\omega)e^{j\omega t}\, d\omega \tag{10.20}$$

If the spectral density of the random noise $v(t)$ is given by $\phi(\omega)$ it follows that

$$Ex_v^2(t) = \frac{1}{2\pi} \int_{-\infty}^{+\infty} |H(j\omega)|^2\, \phi(\omega)\, d\omega \tag{10.21}$$

From (10.18), (10.20), and (10.21) an expression for the output signal-to-noise ratio can be derived.

$$\rho_0 = \frac{1}{2\pi} \frac{\left[\int_{-\infty}^{+\infty} H(j\omega)S(j\omega)e^{j\omega t_0}\, d\omega \right]}{\int_{-\infty}^{+\infty} |H(j\omega)|^2\, \phi(\omega)\, d\omega} \tag{10.22}$$

Applying the Schwarz inequality to the numerator of (10.22) yields

$$\left[\int_{-\infty}^{+\infty} H(j\omega)S(j\omega)e^{j\omega t_0}\, d\omega \right]^2 \leq \int_{-\infty}^{+\infty} |H(j\omega)|^2\, \phi(\omega)\, d\omega \int_{-\infty}^{+\infty} \frac{|e^{j\omega t_0}|^2\, |S(j\omega)|^2}{\phi(\omega)}\, d\omega \tag{10.23}$$

Consequently, from (10.22) and (10.23) and since $|e^{\omega j t_0}| = 1$ we have

$$\rho_0 \leq \frac{1}{2\pi} \int_{-\infty}^{+\infty} \frac{|S(j\omega)|^2}{\phi(\omega)}\, d\omega \tag{10.24}$$

Hence (10.24) yields an upper bound for the output signal-to-noise ratio. From (10.23) this upper bound can be achieved if

$$H(j\omega) = \alpha e^{-j\omega t_0} \frac{S(-j\omega)}{\phi(\omega)} \tag{10.25}$$

where α is an arbitrary constant. This statement can be verified by directly substituting for $H(j\omega)$ in (10.23) and observing that the inequality is then satisfied as equality. The optimal (maximum) value of the output signal-to-noise ratio is given by

$$\rho_{\max} = \max_{H(j\omega)} \rho_0 = \frac{1}{2\pi} \int_{-\infty}^{+\infty} \frac{|S(j\omega)|^2}{\phi(\omega)}\, d\omega \tag{10.26}$$

Equation (10.25) yields the best possible linear filter with the objective of maximizing the value of the output signal-to-noise ratio at a given time t_0.

It is interesting to derive the values of $x_s^2(t_0)$ and $Ex_v^2(t)$ when the filter given by (10.25) is used. From (10.20) and (10.25) we have

$$x_s(t_0) = \frac{\alpha}{2\pi} \int_{-\infty}^{+\infty} \frac{|S(j\omega)|^2}{\phi(\omega)}\, d\omega \tag{10.27}$$

Hence, using (10.27) and $\alpha = 1$, we obtain

$$x_s(t_0) = \rho_{\max} \quad \text{and} \quad x_s^2(t_0) = \rho_{\max}^2$$

From (10.21) and also for $\alpha = 1$, it follows that

$$Ex_v^2(t) = \frac{1}{2\pi} \int_{-\infty}^{+\infty} \frac{|S(j\omega)|^2}{\phi(\omega)}\, d\omega = \rho_{\max} \tag{10.28}$$

which yields

$$\frac{S_0}{N_0} = \frac{x_s^2(t_0)}{Ex_v^2(t)} = \rho_{\max} \qquad (10.29)$$

It should be observed that the maximum value of signal-to-noise ratio at the output of the filter is independent of α as long as α is different from zero. This, from (10.25), is equivalent to saying that the optimal filter is independent of signal amplitude. Hence, the filter remains the same even if the signal amplitude is known (i.e., $s(t)$ can be substituted by $\beta s(t)$ where β is unknown). By the same token, the optimality of the optimal filter is not affected by a multiplicative gain constant. Thus, if $H(j\omega)$ is the optimal filter, then $\alpha H(j\omega)$ is an equally good optimal filter.

On the other hand, if the signal has an unknown delayed arrival time, represented by $s(t - \tau)$, use of the filter given by (10.25) will yield a smaller value for signal-to-noise ratio at t_0 since the maximum now occurs at $t_0 + \tau$. However, since τ is unknown we are forced to choose the same filter as before.

10.5 The Matched Filter

Let us consider the observation noise $v(t)$ in the preceding analysis to be white. Consequently,

$$\phi(\omega) = 1$$

and (10.25) becomes (for $\alpha = 1$)

$$H(j\omega) = e^{-j\omega t_0} S(-j\omega) \qquad (10.30)$$

This filter is called the matched filter. Its transfer function is completely determined from the knowledge of the signal waveform $s(t)$. The response of this filter to the signal $s(t)$ is given by

$$X_s(j\omega) = e^{-j\omega t_0} S(-j\omega) S(j\omega) \qquad (10.31)$$

or

$$x_s(t) = \frac{1}{2\pi} \int_{-\infty}^{+\infty} S(j\omega) S(-j\omega) e^{-j\omega(t_0-t)} \, d\omega \qquad (10.32)$$

Substituting for $S(\omega)$ from

$$S(\omega) = \int_{-\infty}^{+\infty} s(\tau) e^{-j\omega\tau} \, d\tau \qquad (10.33)$$

into (10.32) yields

$$x_s(t) = \frac{1}{2\pi} \int_{-\infty}^{+\infty} S(-j\omega) e^{-j\omega(t_0-t)} \int_{-\infty}^{+\infty} s(\tau) e^{-j\omega\tau} \, d\tau \, d\omega \qquad (10.34)$$

Interchanging the order of integration results in

$$x_s(t) = \frac{1}{2\pi} \int_{-\infty}^{+\infty} s(\tau) \int_{-\infty}^{+\infty} S(-j\omega)e^{-j\omega(t_0+\tau-t)}\,d\omega\,d\tau \tag{10.35}$$

Since

$$\frac{1}{2\pi} \int_{-\infty}^{+\infty} S(-j\omega)e^{-j\omega(t_0+\tau-t)}\,d\omega = s(t_0 + \tau - t) \tag{10.36}$$

Equation (10.35) becomes

$$x_s(t) = \int_{-\infty}^{+\infty} s(\tau)s(t_0 + \tau - t)\,d\tau \tag{10.37}$$

or finally

$$x_s(t_0) = \int_{-\infty}^{+\infty} s^2(\tau)\,d\tau \tag{10.38}$$

Equation (10.38) indicates that the matched filter at $t = t_0$ produces an output (in response to the input signal) equal (for $\alpha = 1$) to the signal energy.

An interesting property of the matched filter is revealed if we determine its impulse response, denoted by $h(t)$. From (10.30) it follows that

$$h(t) = \frac{1}{2\pi} \int_{-\infty}^{+\infty} e^{-j\omega t_0}e^{+j\omega t}S(-j\omega)\,d\omega \tag{10.39}$$

using the change of variable ω to $-\omega$ and the fact that

$$s(t) = \frac{1}{2\pi} \int_{-\infty}^{+\infty} S(j\omega)e^{j\omega t}\,d\omega \tag{10.40}$$

we get

$$h(t) = s(t_0 - t) \tag{10.41}$$

Since the output $x(t)$ and input $y(t)$ are related by the convolution integral

$$x(t) = \int_{-\infty}^{+\infty} h(t - \tau)y(\tau)\,d\tau \tag{10.42}$$

substituting for $h(t)$ from (10.41) yields

$$x(t) = \int_{-\infty}^{+\infty} s(t_0 - t + \tau)y(\tau)\,d\tau \tag{10.43}$$

The output at $t = t_0$ is then given by

$$x(t_0) = \int_{-\infty}^{+\infty} s(t_0 - t_0 + \tau)y(\tau)\,d\tau \tag{10.44}$$

More specifically, if the signal $s(t)$ is zero outside the interval $0 \leq t \leq T$, then (10.44) becomes

$$x(t_0) = \int_0^T s(t)y(t)\,dt \tag{10.45}$$

Equations (10.43) and (10.45) reveal that the action of the matched filter is to cross-correlate the signal $s(t)$ with the observation. This result accomplishes the desired operation given by (10.12) and (10.13) of the preceding section. It is interesting that the matched filter, being a linear filter, has accomplished this purpose.

▶ **example**

Let the signal be given by

$$s(t) = 1 \qquad 0 \leq t \leq T$$
$$\quad = 0 \qquad t < 0, t > T \tag{10.46}$$

and the observation by

$$y(t) = s(t) + v(t)$$

where $v(t)$, the observation noise, is white. It is desired to find a filter producing maximum signal-to-noise ratio at $t = T$. According to (10.45) the optimal filter is an integrator which initiates the integration of $y(t)$ at $t = 0$ and terminates at $t = T$. The optimal transfer function is derived from (10.30) where

$$S(j\omega) = \int_0^T e^{-j\omega t}\, dt \tag{10.47}$$

Hence

$$H(j\omega) = \int_0^T e^{j\omega(t-T)}\, dt \tag{10.48}$$

Performing the integration results in

$$H(j\omega) = \frac{1 - e^{-j\omega T}}{j\omega}$$

This is the transfer function of an integrator with an integration interval $[0, T]$, which can be written alternatively as

▶
$$H(j\omega) = Te^{-j\omega T/2} \frac{\sin \omega T/2}{\omega T/2}$$

PROBLEMS

1. Let the signal $s(t)$ be given by

$$s(t) = t \qquad 0 \leq t \leq T$$
$$\quad = 0 \qquad t < 0, t > T$$

and the observation by

$$y(t) = s(t) + v(t)$$

where $v(t)$ is the noise term and is an independent process. Derive a linear filter operating on $y(t)$ which maximizes the signal-to-noise ratio at time T.

2. Determine the filter which maximizes the signal-to-noise ratio at $t = \pi/\omega_0$ for the signal $s(t)$ given by

$$s(t) = a \sin^2 \omega_0 t$$

and noise $n(t)$ with autocorrelation function

$$\varphi_n(\tau) = \frac{1}{\alpha} e^{-\alpha|\tau|}$$

α, a, and ω_0 are constants.

3. In Problem 2, let $\varphi_n(\tau) = \delta(\tau)$ and derive the corresponding matched filter.

SELECTED READINGS

Davenport, W. B., Jr. and W. L. Root, *An Introduction to the Theory of Random Signals and Noise*, McGraw-Hill, 1958.

Papoulis, A., *Probability, Random Variables, and Stochastic Processes*, McGraw-Hill, New York, 1965.

Van Vleck, J. H. and D. Middleton, "A Theoretical Comparison of Visual, Aural, and Meter Reception of Pulsed Signals in the Presence of Noise," *J. Applied Physics*, November 1946.

Wainstein, L. A. and V. D. Zubakov, *Extraction of Signals From Noise*, translation from Russian, Prentice-Hall, Englewood Cliffs, N.J., 1962.

Zadeh, L. A. and J. R. Ragazini, "Optimum Filters for the Detection of Signals in Noise," *Proc. IRE*, October 1952.

chapter 11 / CRAMÉR-RAO LOWER BOUND AND ITS APPLICATIONS

11.1 Introduction

Let us consider the problem of estimating a scalar x by processing the observation $y(i) = f[x, v(i), i]$, $1 \leq i \leq k$, where $v(i)$ is the observation noise with a known probability density function. In the preceding chapters, various optimal estimators were derived, each of which yielded an optimal estimate $\hat{x}$ of x which satisfied a particular criterion of optimality. It is customary (and usually convenient) to choose, rather arbitrarily, the conditional or unconditional expected square error of estimation, namely $E_v[(x - \hat{x}(y))^2 \mid x]$ or $E_x E_v[(x - \hat{x}(y))^2]$ as a universal measure of the "quality" of all estimators. Here E_v and E_x refer to expectation over the noise v and parameter x, respectively,* and $\hat{x}$ represents the estimator, the quality of which we wish to establish. We know that $\hat{x}$, in general, is a function of $y(i)$, and consequently is a random variable. By carrying out the indicated expectations for various estimators, we may arrive at different values for estimation error variance. Unfortunately, the expectation operation leading to this measure is, in general, very complicated, owing to the complexity of various estimators. However, it is possible to derive an expression for a lower bound on the variance in terms of only the statistical properties of the

---

* Clearly the latter would be meaningful if $p(x)$ were known.

246

observed signal and the estimator bias. This is a very interesting result, since we can establish at least a lower bound on the quality of any estimator without having any knowledge of the estimator itself except, for example, that it is unbiased. The derivation of this important result and its various uses in estimation theory is the subject of this chapter.

11.2 Derivation of the Lower Bound

Let $\hat{x}$ be an estimator of x, i.e., $\hat{x}$ represents a function of the observations $y(i)$, $i = 1, \ldots, k$ where

$$y(i) = f[x, v(i), i] \qquad i = 1, \ldots, k \tag{11.1}$$

and $v(i)$ is the observation noise with known statistics, so that the joint conditional probability density function $p[y(1), \ldots, y(k) \mid x]$ is known. Let us denote the bias of the estimator $\hat{x}$ by $\varphi(x)$. Hence, considering the fact that $\hat{x}$ as an estimator is a random variable owing to its dependence on $y(i)$, we have

$$E[\hat{x} \mid x] \overset{\triangle}{=} \psi(x) = x + \varphi(x) \tag{11.2}$$

and

$$\psi(x) \overset{\triangle}{=} \int_{\hat{x}} \hat{x} p(\hat{x} \mid x) \, d\hat{x} \tag{11.3}$$

where $p(\hat{x} \mid x)$, the probability density of $\hat{x}$ given x, is completely determined by the form of dependence of $\hat{x}$ on $y(i)$, equation (11.1), and the statistics of the observation noise $v(i)$. The integration in (11.3) is over all possible values of $\hat{x}$. Since $p(\hat{x} \mid x)$ is, for any x, a probability density on $\hat{x}$, we clearly have

$$\int_{\hat{x}} p(\hat{x} \mid x) \, d\hat{x} = 1 \tag{11.4}$$

where x appears as a constant parameter in (11.4). Differentiating (11.4) with respect to this parameter yields

$$\int_{\hat{x}} \frac{\partial p(\hat{x} \mid x)}{\partial x} \, d\hat{x} = 0 \tag{11.5}$$

Since the integration in (11.5) is with respect to $\hat{x}$, we can multiply both sides by x. It follows that

$$\int_{\hat{x}} x \frac{\partial p(\hat{x} \mid x)}{\partial x} \, d\hat{x} = 0 \tag{11.6}$$

Differentiating both sides of (11.3) with respect to x yields the identity

$$\frac{\partial \psi(x)}{\partial x} = \int_{\hat{x}} \hat{x} \frac{\partial p(\hat{x} \mid x)}{\partial x} \, d\hat{x} \tag{11.7}$$

Subtracting (11.6) from the right hand side of (11.7), it follows that

$$\frac{\partial \psi(x)}{\partial x} = \int_{\hat{x}} (\hat{x} - x) \frac{\partial p(\hat{x} \mid x)}{\partial x} \, d\hat{x} \tag{11.8}$$

Multiplying and dividing the integrand of (11.8) by $\sqrt{p(\hat{x} \mid x)}$, we have

$$\frac{\partial \psi(x)}{\partial x} = \int_{\hat{x}} (\hat{x} - x) \sqrt{p(\hat{x} \mid x)} \, \frac{\dfrac{\partial p(\hat{x} \mid x)}{\partial x}}{\sqrt{p(\hat{x} \mid x)}} \, d\hat{x}$$

Taking the absolute value of both sides and applying the Schwarz inequality (for integrals) yields

$$\left[\frac{\partial \psi(x)}{\partial x} \right]^2 \leq \left\{ \int_{\hat{x}} (\hat{x} - x)^2 p(\hat{x} \mid x) \, d\hat{x} \right\} \left\{ \int_{\hat{x}} \left[\frac{\partial p(\hat{x} \mid x)}{\partial x} \right]^2 \frac{1}{p(\hat{x} \mid x)} \, d\hat{x} \right\} \tag{11.9}$$

Consequently, identifying the first integral on the right hand side of (11.9) as the expected square error of estimation, we obtain

$$E[(\hat{x} - x)^2 \mid x] \geq \frac{\left[\dfrac{\partial \psi(x)}{\partial x} \right]^2}{\displaystyle\int_{x} \left[\frac{\partial p(\hat{x} \mid x)}{\partial x} \right]^2 \frac{1}{p(\hat{x} \mid x)} \, d\hat{x}} \tag{11.10}$$

Equation (11.10) yields a lower bound for the estimation error variance conditioned on x. However, it is not in a desirable form since, for its evaluation, we need the density function $p(\hat{x} \mid x)$ which depends on the form of the estimator. In the following, an upper bound for the denominator of (11.10) is derived in terms of the density $p(y \mid x)$. This conditional density function does not depend on the choice of the estimator.

Let the joint conditional density function of the observation vector y be denoted by $p(y \mid x)$ where y is a k-dimensional vector with elements $y(1), \ldots,$ $y(k)$. Let us transform the k-dimensional vector y into a k-dimensional vector z where $z = z(y)$ and $y = f(z)$ are known to exist. Therefore, the conditional density $p(z \mid x)$ is given by

$$p(z \mid x) = p[y = f(z) \mid x] \, |J| \tag{11.11}$$

where $|J|$ is the Jacobian of transformation $y = f(z)$ and is independent of x. Let us substitute for $p(z \mid x)$ in the following expression

$$\int_{z} \left[\frac{\partial p(z \mid x)}{\partial x} \right]^2 \frac{1}{p(z \mid x)} \, dz$$

where we also have

$$dz\,|J| = dy$$

It follows that

$$\int_z \left[\frac{\partial p(z\mid x)}{\partial x}\right]^2 \frac{1}{p(z\mid x)}\,dz = \int_y \left[\frac{\partial p(y\mid x)}{\partial x}\right]^2 \frac{1}{p(y\mid x)}\,dy \qquad (11.12)$$

Now let the function $z(y)$ be chosen such that the first component of $z(y)$ is $\hat{x}$, i.e.,

$$z(y) = \begin{bmatrix} \hat{x}(y) \\ \hline \xi(y) \end{bmatrix} \qquad (11.13)$$

where $\hat{x}(y)$ is scalar and $\xi(y)$ is an $k-1$ dimensional vector. Consequently, $p(z\mid x)$ can be written

$$p(z\mid x) = p[\hat{x}, \xi \mid x] = p(\hat{x}\mid x)p(\xi\mid \hat{x}, x) \qquad (11.14)$$

Therefore,

$$\frac{\partial p(z\mid x)}{\partial x} = \frac{\partial p(\hat{x}\mid x)}{\partial x}p(\xi\mid \hat{x}, x) + p(\hat{x}\mid x)\frac{\partial p(\xi\mid \hat{x}, x)}{\partial x} \qquad (11.15)$$

Substituting (11.15) into (11.12) yields

$$\int_y \left[\frac{\partial p(y\mid x)}{\partial x}\right]^2 \frac{1}{p(y\mid x)}\,dy = \int_{\hat{x},\xi} \left[\frac{\partial p(\hat{x}\mid x)}{\partial x}\right]^2 \frac{p(\xi\mid \hat{x}, x)}{p(\hat{x}\mid x)}\,d\hat{x}\,d\xi$$

$$+ \int_{\hat{x},\xi} \left[\frac{\partial p(\xi\mid \hat{x}, x)}{\partial x}\right]^2 \frac{p(\hat{x}\mid x)}{p(\xi\mid \hat{x}, x)}\,d\hat{x}\,d\xi$$

$$+ 2\int_{\hat{x},\xi} \frac{\partial p(\hat{x}\mid x)}{\partial x}\frac{\partial p(\xi\mid \hat{x}, x)}{\partial x}\,d\hat{x}\,d\xi \qquad (11.16)$$

From (11.5) and

$$\int_\xi p(\xi\mid \hat{x}, x)\,d\xi = 1$$

and the fact that the second term in the right hand side of (11.6) is always positive, we obtain

$$\int_y \left[\frac{\partial p(y\mid x)}{\partial x}\right]^2 \frac{1}{p(y\mid x)}\,dy \geq \int_{\hat{x}} \left[\frac{\partial p(\hat{x}\mid x)}{\partial x}\right]^2 \frac{1}{p(\hat{x}\mid x)}\,d\hat{x} \qquad (11.17)$$

From (11.10) and (11.17) it follows that

$$E[(\hat{x} - x)^2 \mid x] \geq \frac{\left[\dfrac{\partial \psi(x)}{\partial x}\right]^2}{\displaystyle\int_y \left[\dfrac{\partial p(y\mid x)}{\partial x}\right]^2 \dfrac{1}{p(y\mid x)}\,dy} \qquad (11.18)$$

Equation (11.18) is the well-known Cramér-Rao inequality. This inequality establishes a lower bound on the expected square estimation error in terms of the statistics of the observation, $p(y \mid x)$, and the estimator bias $\varphi(x)$, where

$$\frac{\partial \psi(x)}{\partial x} = 1 + \frac{\partial \varphi(x)}{\partial x}$$

The denominator of the right hand side of (11.18) can be written in any of the following equivalent forms:

$$\int_y \left[\frac{\partial p(y \mid x)}{\partial x} \right]^2 \frac{1}{p(y \mid x)} \, dy = \int_y \left[\frac{\partial \log p(y \mid x)}{\partial x} \right]^2 p(y \mid x) \, dy$$

$$= E\left[\frac{\partial \log p(y \mid x)}{\partial x} \right]^2 \qquad (11.19)$$

If the estimator is unbiased, i.e., if $\varphi(x)$ in (11.2) is identically zero, we have

$$\frac{\partial \psi(x)}{\partial x} = 1$$

and (11.18) becomes

$$E[(\hat{x} - x)^2 \mid x] \geq \frac{1}{\int_y \left[\dfrac{\partial p(y \mid x)}{\partial x} \right]^2 \dfrac{1}{p(y \mid x)} \, dy} \qquad (11.20)$$

When the observations $y(1), \ldots, y(k)$ are statistically independent it follows that

$$p(y \mid x) = p[y(1), y(2), \ldots, y(k) \mid x] = \prod_{i=1}^{k} p[y(i) \mid x]$$

Therefore

$$\frac{\partial p(y \mid x)}{\partial x} = \sum_{j=1}^{k} \frac{\partial p[y(j) \mid x]}{\partial x} \prod_{\substack{i=1 \\ i \neq j}}^{k} p[y(i) \mid x]$$

Hence,

$$\int_y \left[\frac{\partial p(y \mid x)}{\partial x} \right]^2 \frac{1}{p(y \mid x)} \, dy = \int_y \sum_{i=1}^{k} \left[\frac{\partial p[y(j) \mid x]}{\partial x} \right]^2 \frac{1}{p[y(j) \mid x]} \, dy \qquad (11.21)$$

If the function f in (11.1) is not an explicit function of i, and if the noise sequence $v(i)$ is stationary, then the integrals of each term of the summation in (11.21) are identical. Consequently, (11.18) assumes the form

$$E[(\hat{x} - x)^2 \mid x] \geq \frac{\left[\dfrac{\partial \psi(x)}{\partial x} \right]^2}{k \int_{y(i)} \left[\dfrac{\partial p(y(i) \mid x)}{\partial x} \right]^2 \dfrac{1}{p(y(i) \mid x)} \, dy(i)} \qquad (11.22)$$

▶ **example**

Let $y(i)$ be a scalar gaussian independent random sequence with known variance σ^2 and unknown mean x. We wish to determine a lower bound on the expected square error of any estimate $\hat{x}$ which could conceivably be obtained by processing $y(1), \ldots, y(k)$ in any manner whatsoever. We can write

$$y(i) = x + v(i)$$

$$Ev(i) = 0$$

$$Ev(i)v(j) = \sigma^2 \, \Delta(i - j)$$

Therefore

$$p[y(i) \mid x] = \frac{1}{\sqrt{2\pi}\,\sigma} \exp -\frac{1}{2} \frac{[y(i) - x]^2}{\sigma^2}$$

Hence,

$$\frac{\partial p[y(i) \mid x]}{\partial x} = \frac{1}{\sqrt{2\pi}\,\sigma}\left[\frac{y(i) - x}{\sigma^2}\right] \exp -\frac{1}{2} \frac{[y(i) - x]^2}{\sigma^2}$$

Substituting into (11.22) and restricting our consideration to unbiased estimators, we have

▶
$$E[(x - \hat{x})^2 \mid x] \geq \frac{\sigma^2}{k} \tag{11.23}$$

This example was considered in Section 8.2 where we derived the maximum likelihood estimate of x, which was shown to be the sample mean. It can easily be shown that the sample mean is indeed an unbiased estimator of the mean, and that the resultant expected square estimation error is that given by (11.23). Thus, interestingly, the sample mean is an estimator which achieves the lower bound.

11.3 Efficient Estimators

If an unbiased estimator $\hat{x}$ can be found such that the Cramér-Rao inequality is satisfied as an equality, it is called an "efficient" estimator. For example, the sample mean, which happens to be a maximum likelihood estimator, is also an efficient estimator for the problem considered in the previous section. In general, maximum likelihood estimators are not efficient. However, it is an interesting result that their corresponding expected square estimation error approaches the lower bound as the number of observations tends to infinity.* Hence, maximum likelihood estimators (and other estimators having this property) are called asymptotically efficient.

* Cramer, H., *Mathematical Methods of Statistics*.

It can also be shown, as a special case, that when the observation is linear in x and v, then the maximum likelihood estimate is efficient regardless of the length of the data.

11.4 Additive Gaussian Noise

In general the evaluation of the integrals in the expressions (11.18), (11.20), or (11.21) is exceedingly complicated. However, the integrals assume a very simple form if (11.1) can be written as

$$y(i) = h(x, i) + v(i), \qquad i = 1, \ldots, k \tag{11.24}$$

or in vector form

$$y = h(x) + v \tag{11.25}$$

where y, h, and v are k-vectors whose ith components satisfy (11.24), and where v is a gaussian random vector with zero-mean and a k-dimensional covariance matrix given by

$$Evv' = L \tag{11.26}$$

The joint probability density function of y conditioned on the parameter x will clearly be gaussian with mean $h(x)$ and covariance L

$$p(y \mid x) = C \exp -\tfrac{1}{2}[y - h(x)]'L^{-1}[y - h(x)] \tag{11.27}$$

where C is a constant independent of x. Taking the logarithm of both sides of (11.27), it follows that

$$\log p(y \mid x) = \log C - \tfrac{1}{2}[y - h(x)]'L^{-1}[y - h(x)]$$

and, upon differentiation with respect to the scalar parameter x,

$$\frac{\partial \log p(y \mid x)}{\partial x} = [y - h(x)]'L^{-1}\frac{\partial h(x)}{\partial x} \tag{11.28}$$

Since (11.28) is a scalar, and since L is symmetric, then

$$\left[\frac{\partial \log p(y \mid x)}{\partial x}\right]^2 = \left[\frac{\partial h(x)}{\partial x}\right]'L^{-1}[y - h(x)][y - h(x)]'L^{-1}\frac{\partial h(x)}{\partial x} \tag{11.29}$$

Now it follows from (11.24) and (11.26) that

$$E[y - h(x)][y - h(x)]' = Evv' = L$$

Hence, taking the expected value of both sides of (11.29) with respect to y, we obtain

$$E\left[\frac{\partial \log p(y \mid x)}{\partial x}\right]^2 = \left[\frac{\partial h(x)}{\partial x}\right]'L^{-1}\frac{\partial h(x)}{\partial x} \tag{11.30}$$

Substituting into (11.18) and considering (11.19) yields

$$E[(\hat{x} - x)^2 \mid x] \geq \frac{\left[\dfrac{\partial \psi(x)}{\partial x}\right]^2}{\left[\dfrac{\partial h(x)}{\partial x}\right]' L^{-1} \dfrac{\partial h(x)}{\partial x}} \tag{11.31}$$

This is the Cramér-Rao lower bound for the special case of additive gaussian noise. If the gaussian observation noise $v(i)$ is a stationary uncorrelated sequence, so that $L = \sigma^2 I$, (11.30) becomes

$$E\left[\frac{\partial \log p(y \mid x)}{\partial x}\right]^2 = \frac{1}{\sigma^2} \left\| \frac{\partial h(x)}{\partial x} \right\|^2 \tag{11.32}$$

where

$$Ev^2(i) = \sigma^2$$

It is evident that since $h(x)$ is a known function of x, the evaluation of (11.31) [or (11.32)] is in general far easier than (11.18) [or (11.22)].

Similar results can be obtained for the case of continuous observations. The following is a formal derivation. Let

$$y(t) = h(x, t) + v(t) \qquad 0 \leq t \leq T$$
$$Ev(t_1)v(t_2) = l(t, \tau) \tag{11.33}$$

where y, h, v, and l are scalars. Let us consider the equation

$$y(t_i) = h(x, t_i), \qquad i = 1, \ldots, k, \tag{11.34}$$

along with

$$Ev(t_i)v(t_j) = l(i, j)$$

Therefore, from (11.30) we have

$$E\left[\frac{\partial \log p(y \mid x)}{\partial x}\right]^2 = \sum_{j=1}^{k} \sum_{i=1}^{k} \frac{\partial h(x, t_i)}{\partial x} \frac{\partial h(x, t_j)}{\partial x} r(i, j) \tag{11.35}$$

where $r(i, j)$ is the i, j-th element of the *inverse* of the covariance matrix L whose elements are $l(i, j)$, so that

$$\sum_{j=0}^{k} l(i, j) r(j, m) = \Delta(i - m) \tag{11.36}$$

Letting k approach infinity while $t_1 = 0$, $t_k = T$, equation (11.35) becomes, formally,

$$E\left[\frac{\partial \log p(y \mid x)}{\partial x}\right]^2 = \int_0^T \int_0^T \frac{\partial h(x, t)}{\partial x} \frac{\partial h(x, \tau)}{\partial x} r(t, \tau) \, dt \, d\tau \tag{11.37}$$

and (11.36) becomes

$$\int_0^T l(t, \gamma) r(\gamma, \tau) \, d\gamma = \delta(t - \tau) \tag{11.38}$$

The right hand side of (11.37) is then equivalent to

$$\int_0^T \int_0^T \frac{\partial h(x, t)}{\partial x} \frac{\partial h(x, \tau)}{\partial x} r(t, \tau) \, dt \, d\tau = \int_0^T \frac{\partial h(x, t)}{\partial x} \frac{\partial f(x, t)}{\partial x} \, dt \tag{11.39}$$

where $f(x, \gamma)$ satisfies the inhomogeneous Fredholm integral equation*

$$\int_0^T h(x, \tau) r(\gamma, \tau) \, d\tau = f(x, \gamma) \tag{11.40}$$

Multiplying both sides of (11.40) by $l(t, \gamma)$ and integrating over $[0, T]$, it follows that

$$\int_0^T f(x, \gamma) l(t, \gamma) \, d\gamma = \int_0^T \int_0^T h(x, \tau) l(t, \gamma) r(\gamma, \tau) \, d\tau \, d\gamma \tag{11.41}$$

Substituting from (11.38) yields

$$\int_0^T f(x, \gamma) l(t, \gamma) \, d\gamma = h(x, t) \tag{11.42}$$

Finally, from (11.37) and (11.39), the Cramér-Rao inequality takes the form

$$E[(\hat{x} - x)^2 \mid x] \geq \frac{\left[\dfrac{\partial \psi(x)}{\partial x}\right]^2}{\displaystyle\int_0^T \frac{\partial h(x, t)}{\partial x} \frac{\partial f(x, t)}{\partial x} \, dt} \tag{11.43}$$

where $f(x, t)$ is the solution of the integral equation given by (11.42), or (11.40) and (11.38).

For the special case where the additive observation noise $v(t)$ is zero-mean, stationary white (gaussian) with autocorrelation given by

$$l(t, \tau) = l_0 \delta(t - \tau)$$

equation (11.42) yields

$$h(x, t) = l_0 f(x, t)$$

and (11.43) becomes

$$E[(\hat{x} - x)^2 \mid x] \geq \frac{l_0 \left[\dfrac{\partial \psi(x)}{\partial x}\right]^2}{\displaystyle\int_0^T \left[\frac{\partial h(x, t)}{\partial x}\right]^2 \, dt} \tag{11.44}$$

* Helstrom, C. W., Statistical Theory of Signal Detection, Pergamon, New York, 1960.

▶ **example 1**

A vehicle is flying on a straight line course with initial position 1 and unknown initial velocity x. The instantaneous position is measured over the interval $[0, T]$. The observation contains additive zero-mean white gaussian noise. Hence,

$$y(t) = 1 + xt + v(t)$$
$$Ev(t_1)v(t_2) = l_0\delta(t_2 - t_1)$$

Comparing the above expressions with (11.33), we see that the function $h(x, t)$ is given by

$$h(x, t) = 1 + xt$$

Substituting into (11.44) and considering the case of an unbiased estimator, it follows that

$$E[(\hat{x} - x)^2 \mid x] \geq \frac{3l_0}{T^3}$$

where the right-hand side of the above inequality is the Cramér-Rao lower bound.

It can easily be shown that the maximum likelihood estimate of x is given by

$$\hat{x} = \frac{\displaystyle\int_0^T ty(t)\,dt - \frac{T^2}{2}}{\dfrac{T^3}{3}}$$

and

$$E\hat{x} = x$$

$$E[(\hat{x} - x)^2 \mid x] = \frac{3l_0}{T^3}$$

Thus, the maximum likelihood estimator is unbiased, and has the minimum variance property. Therefore, in this case the maximum likelihood ▶ estimator is efficient.

▶ **example 2**

Let the signal $h(x, t)$ be given by

$$h(x, t) = x \sin t$$

and the observation $y(t)$ by

$$y(t) = x \sin t + v(t)$$
$$Ev(t_1)v(t_2) = l_0\delta(t_2 - t_1)$$

Substituting into (11.44) we obtain the following lower bound for the variance of any unbiased estimator of x:

$$E[(\hat{x} - x)^2 \mid x] \geq \frac{l_0}{\displaystyle\int_0^T \sin^2 t \, dt} \tag{11.45}$$

The maximum likelihood estimate of x for this example was obtained in Section 8.3. Comparison of (11.5) with (8.30) reveals that the maximum likelihood estimator achieves the lower bound. Consequently, for this
▶ problem, the maximum likelihood estimator is also an efficient estimator.

▶ **example 3**

In Example 1 of Section 8.7, we derived an unbiased maximum likelihood estimator for x where

$$y(t) = xh(t) + v(t)$$
$$Ev(t_1)v(t_2) = \varphi(t_2 - t_1)$$

From (11.43) we have

$$E[(\hat{x} - x)^2 \mid x] \geq \frac{1}{\displaystyle\int_0^T h(t) \frac{\partial f(x, t)}{\partial x} \, dt}$$

where from (11.42)

$$\int_0^T f(x, \gamma)\varphi(t - \gamma) \, d\gamma = h(t) \cdot x, \qquad (x \text{ is a parameter}).$$

Again, for this example, (8.80) indicates that the maximum likelihood
▶ estimator has achieved the lower bound.

▶ **example 4**

Let us consider a very simple version of a radar range measurement problem. We transmit a waveform described by $h(t)$. The reflected waveform from a point-source target will then be given by

$$h(t - x)$$

where $x = \dfrac{2d}{c}$, d is the distance between the transmitter-receiver and the target, and c is the velocity of light. The parameter x is therefore the delay time in target return, so that estimation of x is equivalent to estimation of the target-transmitter range. Let the observed signal $y(t)$ contain white gaussian noise $v(t)$

$$y(t) = h(t - x) + v(t)$$
$$Ev(t_1)v(t_2) = l_0\delta(t_2 - t_1)$$

Using (11.44), we can establish the best possible performance of any unbiased estimator of x in terms of the Cramér-Rao lower bound on the error variance

$$E[(\hat{x} - x)^2 \mid x] \geq \frac{l_0}{\int_0^T \left[\frac{\partial h(t-x)}{\partial x}\right]^2 dt} = \frac{l_0}{\int_0^T \left[\frac{dh(\tau)}{d\tau}\bigg|_{\tau=t-x}\right]^2 dt}$$

If the observation is received over a large enough interval to capture the reflected wave in its entirety, then

$$E[(\hat{x} - x)^2 \mid x] \geq \frac{l_0}{\int_{-\infty}^{+\infty} \left[\frac{\partial h(t)}{\partial t}\right]^2 dt}$$

This result confirms the intuition that, in the absence of any signal attenu-
▶ ation, the quality of the estimate is independent of x, the true range value.

11.5 Signal Design for Optimal Estimation

One problem of estimation that has concerned us throughout this book has been to operate somehow on the "given" observed signal in order to extract information concerning the unknown parameters. The observed signal has been a known function of the unknown parameters, except for the observation noise. However, in certain applications, we have the capability of changing the form of the observed signal. More precisely, the observed signal may be a function of another independent function, usually called the input, in addition to its dependence on the unknown parameters, observation noise, and time. Denoting this input by $\xi(i)$, we then have

$$y(i) = f[x, v(i), i, \xi(i)]$$

or similarly in the continuous case*

$$y(t) = f[x, v(t), t, \xi(t)]$$

The problem posed here is to choose an admissible function $\xi(i)$ or $\xi(t)$ such that the estimate of x is improved in a certain sense. By *admissible* it is meant that $\xi(i)$ or $\xi(t)$ are chosen to satisfy any physical constraint which limits the type of function which may be considered.

Notice that for any "given" $\xi(i)$ or $\xi(t)$ the observation can be represented by the usual functions $f[x, v(i), i]$ or $f[x, v(t), t]$. Consequently, a theoretically logical procedure is then to derive an estimator, say one minimizing the expected square error (or some other criterion) for each possible choice of

* The term $f[\cdot]$ here is, in general, a functional, since it is a function of another function $\xi(t)$ or $\xi(i)$.

function $\xi(i)$ [or $\xi(t)$], and then to choose that $\xi(i)$, [or $\xi(t)$] which will further improve this optimal value of the estimation criterion. However, except for a few trivial cases, this idea is difficult to implement, since it requires knowledge of the functional dependence of the optimal value of the estimation criterion on $\xi(i)$ (or $\xi(t)$). Furthermore, in order to perform the optimization with respect to $\xi(i)$ or $\xi(t)$, this dependence would need to be given in a reasonably simple analytical form in order to be able to carry out the optimization by practical methods. A brief survey of the various estimation procedures reveals that these demands are beyond our present capability, except for the simple linear estimators discussed in Chapters 4–6. It can easily be shown that, in the case of linear estimators, and if $\xi(t)$ also appears linearly in the system model, its choice will have no effect on the quality of the estimates of the states. The proof of this assertion is left as an exercise.

Since the complications stated above are due to the complexity of the estimators and their corresponding criterion functions, the use of the Cramér-Rao lower bound becomes very appealing for the purpose of choosing $\xi(t)$ (or $\xi(i)$), since we recall that this lower bound does not require the knowledge of the specific estimator to be used. In other words, we may decide to choose $\xi(t)$ or $\xi(i)$ to minimize the lower bound. The question is, in doing so, what have we accomplished? The answer: (a) we have optimized the class of efficient estimators, specifically we have achieved the condition under which the "best" maximum likelihood estimator can be obtained when the observed data are long;* (b) we have assured the best mean square error estimator if we succeed in deriving an estimator which achieves the resultant lower bound; (c) we have obtained the lowest lower bound on estimation error variance considering the choice of the input $\xi(t)$ (or $\xi(i)$).

The following sections are devoted to exploration of this concept. Only the case of scalar parameter x, scalar observation y, and additive gaussian noise will be considered, in order that we may use the simple form of the Cramér-Rao lower bound derived in Section 11.4.

11.6   Problem Formulation (Gaussian Noise)

Let the observed signal $y(t)$ be given by

$$y(t) = h[x, t, \xi(t)] + v(t); \qquad 0 \le t \le T \tag{11.46}$$

where y, x, h, and v are scalars; $v(t)$ is stationary, zero-mean white gaussian observation noise, with

$$Ev(t_1)v(t_2) = l_0\,\delta(t_2 - t_1)$$

---

* This is because the maximum likelihood estimator is asymptotically efficient, that is, it achieves the lower bound as the observed data length tends to infinity.

and $\xi(t)$, the input, belongs to a set denoted by S. Each element of this set $[\xi(t), \in S]$ is called an admissible function and by definition satisfies any applicable physical constraints on $\xi(t)$. Typical constraints on $\xi(t)$ are a bound on amplitude of $|\xi(t)|$ or a bound on the energy contained in $\xi(t)$ over the interval of observation $[0, T]$, namely $\int_0^T \xi^2(t)\,dt$. Considering the case of an unbiased estimator of x, (11.44) yields

$$E[(\hat{x} - x)^2 \mid x] \geq \frac{l_0}{\int_0^T \left\{\frac{\partial h[x, t, \xi(t)]}{\partial x}\right\}^2 dt} \qquad (11.47)$$

This lower bound is minimized by maximizing the denominator in the right hand side of (11.47)

$$\max_{\xi(t)\in S} \int_0^T \left\{\frac{\partial h[x, t, \xi(t)]}{\partial x}\right\}^2 dt \qquad (11.48)$$

Similarly, if the observation noise $v(t)$ is not white, i.e.,

$$Ev(t)v(\tau) = l(t, \tau)$$

then, for the unbiased estimator, (11.43) yields

$$E[(\hat{x} - x)^2 \mid x] \geq \frac{1}{\int_0^T \frac{\partial h(x, t, \xi(t))}{\partial x} \frac{\partial f(x, t, \xi(t))}{\partial x} dt} \qquad (11.49)$$

where

$$\int_0^T f[x, \gamma, \xi(\gamma)] l(t, \gamma)\,d\gamma = h[x, t, \xi(t)] \qquad (11.50)$$

In order to minimize the lower bound given by (11.49), the input $\xi(t)$ must then accomplish the following maximization.

$$\max_{\xi(t)\in S} \int_0^T \frac{\partial h[x, t, \xi(t)]}{\partial x} \frac{\partial f[x, t, \xi(t)]}{\partial x} dt \qquad (11.51)$$

The maximizations indicated in (11.48) or (11.51) lead to a choice of the optimal input $\xi(t)$. These maximizations, which are by no means trivial, will be discussed in the following sections.

11.7 Observation Linear in x

A special case of (11.46) is when h is linear in x. Let

$$y(t) = xh[t, \xi(t)] + v(t)$$
$$Ev(t_1)v(t_2) = l_0\,\delta(t_2 - t_1) \qquad (11.52)$$

The operation (11.48) then becomes

$$\max_{\xi(t)\in S} \int_0^T h^2[t,\, \xi(t)]\, dt \tag{11.53}$$

In many applications,* the term $h[t,\, \xi(t)]$ may be the output of a dynamic system with input $\xi(t)$, e.g.,

$$\dot{z}(t) = Az(t) + B\xi(t) \qquad z(0) = z_0;$$
$$h[t,\, \xi(t)] = Cz(t) \tag{11.54}$$

where z is an n-vector, ξ and h are scalars, A is an $n \times n$ matrix, B is an $n \times 1$ vector, and C is a $1 \times n$ vector. It is intuitively obvious that if $\xi(t)$ is not constrained in any way, then if we choose $\xi(t)$ very large in magnitude over the interval of observation $[0,\, T]$, the term $xh[t,\, \xi(t)]$ in (11.52) will completely obscure the presence of any observation noise, and hence the estimation of x will be improved. This trivial and impractical solution is avoided by the physical constraint on the input $\xi(t)$. Let the elements of S (i.e., the admissible functions $\xi(t)$) satisfy an energy constraint given by

$$\int_0^T \xi^2(t)\, dt \le \rho \tag{11.55}$$

where ρ is a given constant. Using the method of Lagrange multipliers, the maximization in (11.53) subject to the constraint (11.55) can be carried out by performing the following maximization:

$$\max_{\xi(t)} \int_0^T h^2[t,\, \xi(t)] - \lambda \xi^2(t)\, dt \tag{11.56}$$

where $\lambda > 0$ is the Lagrange multiplier. Considering (11.54), the expression (11.56) may be re-written as

$$\max_{\xi(t)} \int_0^T [z'(t)C'Cz(t) - \lambda \xi^2(t)]\, dt \tag{11.57}$$

subject to

$$\dot{z}(t) = Az(t) + B\xi(t) \tag{11.58}$$

The method of solution of the constrained maximization problem posed by (11.57) and (11.58) is discussed in Appendix A and proceeds as follows. The Hamiltonian H is defined by

$$H = p'(t)Az(t) + p'(t)B\xi(t) - z'(t)C'Cz(t) + \lambda \xi^2(t) \tag{11.59}$$

* Chapter 9.

where $p(t)$ is given by

$$\dot{p}(t) = -\frac{\partial H}{\partial z} = -A'p(t) + 2C'Cz(t) \tag{11.60}$$

$$p(T) = 0$$

The function $\xi(t)$ maximizing (11.57) is denoted by $\hat{\xi}(t)$. It is required that $\hat{\xi}(t)$ minimize the Hamiltonian given by (11.59) at every t. Hence, differentiating (11.59) with respect to ξ and equating to zero (to obtain the minimum) yields

$$\hat{\xi}(t) = -\frac{1}{2\lambda} B'p(t) \tag{11.61}$$

Substituting into (11.58) yields

$$\dot{z}(t) = Az(t) - \frac{1}{2\lambda} BB'p(t) \tag{11.62}$$

The set of linear equations given by (11.60) and (11.62), along with the initial conditions $z(0) = z_0$ and terminal conditions $p(T) = 0$, will uniquely determine the solutions $z(t)$ and $p(t)$. Equation (11.61) then yields the desired optimal input $\hat{\xi}(t)$. Clearly, $\hat{\xi}(t)$ will be obtained in terms of λ. The value of λ will then have to be chosen so that (11.55) is satisfied.

▶ example 1
It is desired to derive the input $\xi(t)$ in order to enhance the quality of the maximum likelihood estimate of the gain constant x in a system with transfer function $\dfrac{x}{1 + s}$ and input $\xi(t)$. The output $y(t)$ includes additive white gaussian observation noise. Figure 11.1 represents the process.

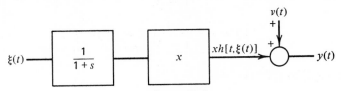

figure 11.1

Hence,

$$y(t) = xh[t, \xi(t)] + v(t) \tag{11.63}$$

$$\dot{z}(t) = -z(t) + \xi(t) \tag{11.64}$$

$$h[t, \xi(t)] = z(t)$$

Equations (11.60) and (11.62) become

$$\dot{z}(t) = -z(t) - \frac{1}{2\lambda} p(t) \tag{11.65}$$

$$\dot{p}(t) = 2z(t) + p(t) \tag{11.66}$$

or equivalently

$$\begin{bmatrix} \dot{z}(t) \\ \dot{p}(t) \end{bmatrix} = \begin{bmatrix} -1 & -\dfrac{1}{2\lambda} \\ 2 & 1 \end{bmatrix} \begin{bmatrix} p(t) \\ z(t) \end{bmatrix} \qquad (11.67)$$

The conditions on $z(t)$ and $p(t)$ are as follows

$$z(0) = z_0$$
$$p(T) = 0 \qquad (11.68)$$

The eigenvalues Λ of the matrix in the right hand side of (11.67) are

$$\pm \sqrt{\frac{\lambda - 1}{\lambda}} \triangleq \pm \Lambda \qquad (11.69)$$

Therefore the solution to $z(t)$ and $p(t)$ for all values of λ except $\lambda = 1$ can be written as*

$$z(t) = C_1 e^{\Lambda t} + C_2 e^{-\Lambda t} \qquad (11.70)$$
$$p(t) = -2\lambda [C_1(1 + \Lambda)e^{\Lambda t} + C_2(1 - \Lambda)e^{-\Lambda t}] \qquad (11.71)$$

where C_1 and C_2 are to be determined by the initial and terminal conditions given by (11.68)

$$z_0 = C_1 + C_2$$
$$0 = C_1(1 + \Lambda)e^{\Lambda T} + C_2(1 - \Lambda)e^{-\Lambda T}$$

Hence,

$$C_1 = \frac{1}{1 - \dfrac{1 + \Lambda}{1 - \Lambda} e^{2\Lambda T}} z(0) \qquad (11.72)$$

$$C_2 = \frac{1}{1 - \dfrac{1 - \Lambda}{1 + \Lambda} e^{-2\Lambda T}} z(0) \qquad (11.73)$$

Substituting into (11.71) yields

$$p(t) = -2\lambda \frac{e^{\Lambda(T-t)} - e^{-\Lambda(T-t)}}{\dfrac{e^{\Lambda T}}{1 - \Lambda} - \dfrac{e^{-\Lambda T}}{1 + \Lambda}} z(0) \qquad (11.74)$$

From (11.61) and (11.74), with $B = 1$, the optimal input is obtained in terms of Λ

$$\hat{\xi}(t) = \frac{e^{\Lambda(T-t)} - e^{-\Lambda(T-t)}}{\dfrac{e^{\Lambda T}}{1 - \Lambda} - \dfrac{e^{-\Lambda T}}{1 + \Lambda}} z(0) \qquad (11.75)$$

* Λ corresponding to $\lambda < 1$ is a pure imaginary complex constant.

The value for Λ is to be chosen so that the inequality constraint (11.55) is satisfied as an equality. This statement is easily justified by noting that if $\xi(t)$ is substituted by $\alpha\xi(t)$, α a positive constant, then the corresponding value of $\int_0^T h^2[t, \xi(t)]\, dt$ will be multiplied by α^2. Consequently if (11.55) is satisfied as an inequality we can choose an $\alpha > 1$ such that $\alpha\xi(t)$ remains admissible, choose $\alpha\xi(t)$ in place of $\xi(t)$, and further increase the value of $\int_0^T h^2[t, \xi(t)]\, dt$.

$$\int_0^T \xi^2(t)\, dt = \frac{1 - \Lambda^2}{2\Lambda} \frac{e^{2\Lambda T} - e^{-2\Lambda T} - 4\Lambda T}{e^{2\Lambda T}\dfrac{1 + \Lambda}{1 - \Lambda} - e^{-2\Lambda T}\dfrac{1 - \Lambda}{1 + \Lambda} - 2} z^2(0) = \rho \quad (11.76)$$

For example let

$$z(0) = 1, \qquad T = 20, \qquad \rho = 1$$

Then by a simple trial and error procedure we obtain

$$\Lambda \approx 0.27$$

And from (11.75)

$$\xi(t) = 0.73 e^{-0.27t}; \qquad 0 \le t \le 20 \tag{11.77}$$

Let us now consider an amplitude constraint on $\xi(t)$ in place of the energy constraint given by (11.55), specifically

$$|\xi(t)| \le \rho \tag{11.78}$$

The conditions that the optimal $\xi(t)$ must satisfy, namely (11.53) and (11.54), are restated below as

$$\max_{\substack{\xi(t) \\ |\xi(t)| \le \rho}} \int_0^T h^2[t, \xi(t)]\, dt \tag{11.79}$$

subject to the constraint

$$\dot{z}(t) = Az(t) + B\xi(t) \tag{11.80}$$

$$h[t, \xi(t)] = Cz(t) \tag{11.81}$$

The solution of the problem posed by (11.79)–(11.81) is also given in Appendix A. The Hamiltonian is defined by

$$H = p'(t)Az(t) + p'(t)B\xi(t) - z'(t)C'Cz(t) \tag{11.82}$$

where $\xi(t)$ must minimize H at every t, $0 \le t \le T$. Hence

$$\xi(t) = -\rho \operatorname{sign} B'p(t) \tag{11.83}$$

where $B'p(t)$ is a scalar and $\operatorname{sign} B'p = 1$ if $B'p > 0$, $\operatorname{sign} B'p = -1$ if

$B'p < 0$. The adjoint vector $p(t)$ satisfies

$$\dot{p}(t) = -\frac{\partial H}{\partial z(t)} = -A'p(t) + 2C'Cz(t) \qquad (11.84)$$

Equations (11.80), (11.83), and (11.84), along with the conditions

$$\begin{aligned} z(0) &= z_0 \\ p(T) &= 0 \end{aligned} \qquad (11.85)$$

will determine the solution $p(t)$. It turns out that the optimal input $\hat{\xi}(t)$ will, in general, assume an amplitude equal to ρ, the maximum allowable value, for all t during the observation interval

▶ **example 2**

Let us consider the same problem as in Example 1 and substitute the constraint

$$|\xi(t)| \leq \rho \qquad (11.86)$$

in place of (11.55). Equations (11.80), (11.83), and (11.84) become

$$\dot{z}(t) = -z(t) - \rho \operatorname{sign} p(t) \qquad (11.87)$$

$$\dot{p}(t) = 2z(t) + p(t) \qquad (11.88)$$

with the initial and terminal conditions

$$\begin{aligned} z(0) &= z_0 \\ p(T) &= 0 \end{aligned} \qquad (11.89)$$

We now have to solve the nonlinear two-point boundary value problem presented by (11.87), (11.88), and (11.89). In general, we must resort to computational procedures. However, for this particular example, we can conjecture a solution $p(t) > 0$, $0 \leq t \leq T$, for $T = 1$ and proceed to verify it. Let us choose

$$z(0) = 0, \qquad \rho = 1, \qquad T = 1$$

Since when $p(t) > 0$

$$\operatorname{sign} p(t) = 1$$

from (11.87) it follows that

$$z(t) = -(1 - e^{-t}) \qquad (11.90)$$

Substituting into (11.88) and solving for $zp(t)$ yields

$$p(t) = e^t p(0) + \int_0^t e^{-s}(e^{-s} - 1)\, ds = e^t p(0) - \tfrac{1}{2}(e^t + e^{-t}) + 1 \qquad (11.91)$$

Satisfying the terminal condition $p(T) = 0$, it follows from (11.91) that

$$p(0) = \frac{\frac{1}{2}(e + e^{-1}) - 1}{e} \tag{11.92}$$

Substituting into (11.91) results in

$$p(t) = [\tfrac{1}{2}e^{-2} - e^{-1}]e^t - \tfrac{1}{2}e^{-t} + 1 \tag{11.93}$$

The conjecture is vindicated, since from (11.93)

$$p(t) > 0 \qquad 0 \le t \le 1$$

Consequently, the optimal input is given by

▶
$$\hat{\xi}(t) = 1 \qquad 0 \le t \le 1 \tag{11.94}$$

11.8 Observation a Nonlinear Function of x

Let a nonlinear dynamic system be governed by the vector differential equation

$$\begin{aligned}\dot{z}(t) &= f[z(t), t, \xi(t), x] \\ z(0) &= z_0\end{aligned} \tag{11.95}$$

where $z(t)$ and f are n-vectors, x is a scalar parameter, and $\xi(t)$ is the scalar input. We are interested in estimating the parameter x by processing the scalar observation $y(t)$ given by

$$y(t) = h[z(t), t] + v(t), \qquad 0 \le t \le T \tag{11.96}$$

where h is a scalar-valued function and $v(t)$ is white gaussian noise. Furthermore, we wish to improve the estimation process by a proper choice of the input $\xi(t)$. Clearly, $h[z(t), t]$ is determined from the knowledge of x and $\xi(t)$. From (11.48), the optimal input $\xi(t)$ must satisfy

$$\max_{\xi(t) \in S} \int_0^T \left[\frac{\partial h[z(t), t]}{\partial x} \right]^2 dt \tag{11.97}$$

Now, we can derive an expression for $\dfrac{\partial h(z(t), t)}{\partial x}$ from (11.95) and (11.96). We have

$$\frac{\partial h[z(t), t]}{\partial x} = \frac{\partial h[z(t), t]'}{\partial z(t)} \frac{\partial z(t)}{\partial x} \tag{11.98}$$

Differentiating both sides of (11.95) with respect to x yields

$$\frac{\partial \dot{z}(t)}{\partial x} = \frac{\partial}{\partial x} f[z(t), t, \xi(t), x] + \left[\frac{\partial f[z(t), t, \xi(t), x]}{\partial z(t)} \right]' \frac{\partial z(t)}{\partial x} \tag{11.99}$$

Changing the order of differentiation with respect to t and x in the left-hand side of (11.95), it follows that

$$\frac{d}{dt}\left[\frac{\partial z(t)}{\partial x}\right] = \frac{\partial f[z(t),\, t,\, \xi(t),\, x]}{\partial x} + \left[\frac{\partial f[z(t),\, t,\, \xi(t),\, x]}{\partial z(t)}\right]' \frac{\partial z(t)}{\partial x} \quad (11.100)$$

Equation (11.100) is a linear vector differential equation in terms of $\partial z(t)/\partial x$, and is called the sensitivity equation. Since $z(0) = z_0$ is independent of x, it follows that

$$\frac{\partial z(t)}{\partial x}\bigg|_{t=0} = 0$$

which provides the initial condition for (11.100). The term

$$\left[\frac{\partial f[z(t),\, t,\, \xi(t),\, x]}{\partial z(t)}\right] \quad (11.101)$$

is an $n \times n$ matrix which, along with the forcing function

$$\frac{\partial f[z(t),\, t,\, \xi(t),\, x]}{\partial x} \quad (11.102)$$

must be evaluated along a solution $z(t)$ for a given value of x. Finally, equations (11.95) and (11.100), for any given x, yield a value for the integrand of (11.97) in terms of $\xi(t)$. In general, the maximization required by (11.97), where the integrand is in terms of the solution of $n + 1$ nonlinear equations [the n-vector equation (11.95) and the scalar equation (11.100) with the input $\xi(t)$], presents a very complicated problem. Various approaches for solving such problems have been suggested by various investigators.* One such approach is by means of the Maximum Principle introduced in Appendix A and already used in the preceding sections. Here, its application will be illustrated by means of an example. It is also observed that the solution $\xi(t)$ is obtained in terms of x. The implication of this is discussed in the next section.

▶ **example**

The problem is to determine the optimal input $\xi(t)$ in order to enhance the estimation of the time constant of a linear first order differential equation. The observed output is corrupted by additive white gaussian noise. The process is modeled by

$$\dot{z}(t) = -xz(t) + x\xi(t) \quad (11.103)$$
$$z(0) = 0$$

$$y(t) = z(t) + v(t) \quad (11.104)$$

* *See* references Athans, Bellman, Bryson, and Kelley at the end of the chapter.

the quantities $\xi(t), z(t), y(t), v(t),$ and x are, respectively, the input, output, observation, observation noise, and inverse of the time constant.* Let the input be bounded in amplitude by unity, i.e.,

$$|\xi(t)| \leq 1 \qquad (11.105)$$

It is evident that if $\xi(t) = 0$, then a meaningful estimation of x is not possible. The question is what input would give us, say, the best maximum likelihood estimate of x.

Let us denote $z(t)$ and $\dfrac{\partial z(t)}{\partial x}$ by $z_1(t)$ and $z_2(t)$, respectively. Equation (11.100) will then become

$$\dot{z}_2(t) = -xz_2(t) - z_1(t) + \xi(t); \qquad z_2(0) = 0 \qquad (11.106)$$

The optimal input $\hat{\xi}(t)$ will then satisfy

$$\max_{\substack{\xi(t) \\ |\xi(t)| \leq 1}} \int_0^t z_2{}^2(t)\, dt \qquad (11.107)$$

Following the procedure outlined in Appendix A, the Hamiltonian H is defined by

$$\begin{aligned} H &= p_1(t)\dot{z}_1(t) + p_2(t)\dot{z}_2(t) - z_2{}^2(t) \\ &= -xp_1(t)z_1(t) + xp_1(t)\xi(t) - xp_2(t)z_2(t) \\ &\quad - p_2(t)z_1(t) + p_2(t)\xi(t) - z_2{}^2(t) \end{aligned} \qquad (11.108)$$

The function $\hat{\xi}(t)$ will minimize H at every t, $0 \leq t \leq T$. Hence

$$\hat{\xi}(t) = -\operatorname{sign}\,[p_1(t) + p_2(t)] \qquad (11.109)$$

The adjoint variables are given by

$$\dot{p}_1 = -\frac{\partial H}{\partial z_1} = +xp_1(t) + p_2(t) \qquad (11.110)$$

$$\dot{p}_2 = -\frac{\partial H}{\partial z_2} = xp_2(t) + 2z_2(t) \qquad (11.111)$$

where

$$p_1(T) = p_2(T) = 0 \qquad (11.112)$$

The four equations (11.103), (11.106), (11.110), and (11.111), along with (11.109) and

$$z_1(0) = z_2(0) = p_1(T) = p_2(T) = 0$$

* The transfer function for the system is clearly $\dfrac{1}{1 + \dfrac{1}{x}s}$.

constitute a two-point boundary-value problem whose solution yields the desired function $\xi(t)$. Since in general the solution can only be obtained by means of computational procedures, we have to choose a value for x and then proceed with computation of $\xi(t)$. For example, for $x = 1$, the following results are obtained

$$\begin{aligned}
\text{if } T \le 2 \quad &\text{then } \hat{\xi}(t) = \pm 1 \quad & 0 \le t \le 2 \\
\text{if } T = 9 \quad &\text{then } \hat{\xi}(t) = \pm 1 \quad & 0 \le t \le 2.8 \\
& \quad\quad\quad = \mp 1 \quad & 2.8 \le t \le 6.5 \\
& \quad\quad\quad = \pm 1 \quad & 6.5 \le t \le 9
\end{aligned}$$

These results are intuitively acceptable, since in the first place it does seem logical to use as large an input as possible $(|\xi(t)| = 1)$. This tends to help diminish the effect of noise (see (11.104)). Furthermore, only the transient portion of $z(t)$ in response to step input is a function of x, and consequently it seems logical that we may wish to apply a new step function (i.e., reverse the input polarity) when the solution tends toward the steady state response. This is indicated above when $T = 9$ by the step change in $\hat{\xi}(t)$ at $t = 2.8$ ▶ and $t = 6.5$.

11.9 Dependence of Input on the Estimation Parameter

In the preceding section, where the observation was assumed nonlinear in x (the parameter to be estimated), it was shown that the input minimizing the Cramér-Rao lower bound was obtained in terms of x. Of course, we do not have an exact value for x, because otherwise there would be no estimation problem at hand. However, we still have to decide on an input for the purpose of estimating x. For this purpose any of the following procedures can be used.

1. Choose the best *a priori* estimate of x, if available, and generate the corresponding $\hat{\xi}(t)$. If the estimation of x can be repeated, this operation will assume a feedback nature, i.e., every time an estimate of x is obtained, it is used to design an input to help improve the next estimate, and so on. (Note that such an estimation process may or may not converge.)
2. If an *a priori* probability density function of x is available, and if it indicates that x is confined within a finite range, then we may choose $\hat{\xi}(t)$ such that it improves the worst possible case, i.e.,

$$\min_{\xi(t)} \max_{x} E[(x - \hat{x})^2 \mid x]$$

This operation may require an exhaustive search over all possible values of x. Consequently, the set of all possible values of x may have to be assumed (or approximated) finite. We can also use $\xi(t)$ minimizing the average value of the conditional expected square error, namely, the unconditional variance

$$\min_{\hat{\xi}(t)} E_x E_y[(x - \hat{x}(y))^2 \mid x] = \min_{\hat{\xi}(t)} E(x - \hat{x}(y))^2$$

Again this requires exhaustion of all possible values of x, since we can only carry out the minimization for a given x, and hence requires that we only consider a finite number of possibilities for values of x.

3. The optimal input $\hat{\xi}(t)$ may turn out to be insensitive to x over its possible range of values. Section 11.7 revealed that the optimal input is in fact independent of x when the observation is linear in x. Intuitively then, $\hat{\xi}(t)$ will not, in general, be very sensitive to the choice of x if the nature of dependence of the observation on x is "close" to linear.

PROBLEMS

1. In the problem statement of Section 4.2 (page 108) replace (4.1) by

$$x(k + 1) = Ax(k) + Bu(k) + F\xi(k)$$

Where F is an $n \times p$ matrix and $\xi(k)$ is the p dimensional input. Given $\xi(k)$ determine an estimator for $x(k + 1)$ minimizing the indicated cost. Show that the quality of the estimator is independent of any choice of $\xi(k)$.

2. In Example 1, Section 8.2, determine whether the estimators $\hat{x}_1$ and $\hat{x}_2$ are efficient.

3. Show whether the maximum likelihood estimate of the amplitude a in the observed signal $y(i)$ is an efficient estimate, where

$$y(i) = a \sin i + v(i)$$

and $v(i)$ is a gaussian random sequence which has the properties

$$Ev(i) = 0$$

$$Ev(i)v(j) = \Delta(j - i)$$

4. Show whether the maximum likelihood estimator derived in Problem (3) of Chapter 8 is efficient.

5. We are given a system represented by Figure 11.2.

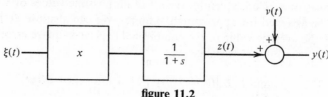

figure 11.2

where $\xi(t)$ is the input, $y(t)$ is the output, and $v(t)$ is white gaussian noise. Derive the input $\xi(t)$, $\int_0^T \xi^2(t)\, dt \leq 1$, such that the lower bound on $E[(x - \hat{x})^2 \mid x]$ is minimized for $z(0) = 0$ and $z(0) = 1$. $\hat{x}$ represents an arbitrary estimator of x and is a function of y. Notice that this problem is different from that of Example 1, Section 11.7, since the sensitivity equation is independent of both the condition $z(0)$ and the parameter value x.

6. Consider Example 1 of Section 11.7. For $z(0) = 0$, show that the optimal input has the form

$$\hat{\xi}(t) = a \sin \omega(T - t)$$

where

$$\tan(\omega T) - \omega = 0$$

SELECTED READINGS

Athans, M. and P. L. Falb, *Optimal Control*, McGraw-Hill, New York, 1966.

Barankin, E. W., "Locally Best Unbiased Estimate," *Annals Math. Statistics*, **20**, 477–501 (1949).

Bellman, R., *Dynamic Programming*, Princeton University Press, Princeton, N.J., 1957.

Bryson, A. E. and W. F. Denham, "A Steepest-Ascent Method for Solving Optimum Programming Problems," *J. Appl. Mech.* **29**, 1962.

Cramér, H., *Mathematical Methods of Statistics*, Princeton University Press, Princeton, N.J., 1946.

Gagliardi, R. M., "Input Selection for Parameter Identification," *Proceedings of the National Electronics Conference* **22**, 1966.

Gart, J. J., "An Extension of Cramér-Rao Inequality," *Annals Math. Statistics*, **30**, No. 2, June, 1959.

Kelley, H. J., "Guidance Theory and Extremal Fields," *IRE Transactions on Automatic Control* **AC-7**, 1962.

Levadi, V. S., "Design of Input Signals for Parameter Estimation," *IEEE Transactions on Automatic Control* **AC-11**, April 1966.

Middleton, D., *An Introduction to Statistical Communication Theory*, McGraw-Hill, New York, 1960.

Nahi, N. E. and D. E. Wallis, Jr., "Optimal Control for Information Maximization in Least-Square Parameter Estimation," University of Southern California, Dept. of Electrical Engineering, 1968, USCEE Report 253.

Rao, C. R., "Information and the Accuracy Attainable in the Estimation of
Statistical Parameters," *Bull. Calcutta Math. Soc.*, 1945.

Slepian, D., "Estimation of Signal Parameters in the Presence of Noise," *IRE
Transactions on Information Theory*, March 1954, V.IT3, pp 68–69.

Wallis, D. E., Jr., "Optimal Input for Information Maximization in Least-Square
Parameter Estimation," doctoral thesis, Department of Electrical Engineering,
University of Southern California, June, 1968.

appendix A

The constrained maximization problem introduced in Chapter 11 in general takes the following form:

Given a vector differential equation

$$\dot{z}(t) = f[z(t), \xi(t), t], \qquad z(0) = z_0 \tag{A.1}$$

and a functional $J[\xi(t)]$ defined by

$$J[\xi(t)] = \int_0^T L[z(t), \xi(t), t] \, dt \tag{A.2}$$

and a set S representing any desired constraint on values of $\xi(t)$. We wish to determine the "admissible" function $\hat{\xi}(t)$ (i.e., one with values belonging to set S) which will yield the maximum of the functional $J[\xi(t)]$

$$\max_{\xi(t) \in S} J[\xi(t)] \tag{A.3}$$

subject to the differential constraint given by (A.1). The functions $z(t)$, $\xi(t)$ are of n, r dimensions respectively and $J[\xi(t)]$ is a scalar.

The above problem is solved by the application of Pontryagin's maximum principle. Let us introduce a scalar function $H[z, \xi(t), t]$ called Hamiltonian as follows

$$H[z, \xi(t), t] \overset{\triangle}{=} p'(t) f[z(t), \xi(t), t] - L[z(t), \xi(t), t] \tag{A.4}$$

where $p(t)$ is an n-dimensional vector called the adjoint vector and defined by the following differential equation

$$\dot{p}(t) \overset{\triangle}{=} - \frac{\partial H[z(t), \xi(t), t]}{\partial z(t)} ; \qquad p(T) = 0 \tag{A.5}$$

273

Pontryagin's Maximum Principle as applied to the above problem is stated as follows:

A necessary condition for $\xi(t) \in S$ to maximize $J[\xi(t)]$ subject to the constraint presented by (A.1) is that $\xi(t)$ minimize the Hamiltonian (A.4) for all t in the interval $[0, T]$.

Proof of this result is found in one of the references* of Chapter 11 and will not be given here. In the following its application is illustrated by means of an example

▶ **example**

Let $z(t)$ and $\xi(t)$ be scalars and

$$\dot{z}(t) = 2z(t) + \xi(t), \qquad z(0) = z_0 \tag{A.6}$$

we wish to maximize over all possible functions $\xi(t)$ the functional

$$J[\xi(t)] = \int_0^T [z^2(t) - \xi^2(t)] \, dt \tag{A.7}$$

The Hamiltonian takes the form

$$H[z(t), \xi(t), t] = 2p(t)z(t) + p(t)\xi(t) + \xi^2(t) - z^2(t) \tag{A.8}$$

Maximizing the Hamiltonian H by differentiating with respect to $\xi(t)$ and equating the results to zero yields

$$\frac{\partial H}{\partial \xi(t)} = p(t) + 2\xi(t) = 0 \tag{A.9}$$

or

$$\xi(t) = -\tfrac{1}{2}p(t) \tag{A.10}$$

The adjoint equation from (A.5) and (A.8) becomes

$$\dot{p}(t) = -\frac{\partial H}{\partial \xi(t)} = -2p(t) + 2z(t) \tag{A.11}$$

$$p(T) = 0$$

Substituting (A.10) into (A.6) yields

$$\dot{z}(t) = 2z(t) - \tfrac{1}{2}p(t) \tag{A.12}$$

$$z(0) = z_0$$

*Athans, M., and P. L. Falb, *Optimal Control*, McGraw-Hill, New York, 1966.

The pair of simultaneous linear equations (A.11) and (A.12) with the indicated boundary conditions can now be solved uniquely for $p(t)$ and $z(t)$. Let us introduce the 2-vector $y(t)$ defined by

$$y(t) = \begin{bmatrix} p(t) \\ z(t) \end{bmatrix}$$

Then we have

$$\dot{y}(t) = Ay(t) \tag{A.13}$$

where

$$A = \begin{bmatrix} -2 & 2 \\ -\frac{1}{2} & 2 \end{bmatrix}; \quad y(0) = \begin{bmatrix} p(0) \\ z_0 \end{bmatrix}; \quad y(T) = \begin{bmatrix} 0 \\ z(T) \end{bmatrix}$$

Equation (A.13) yields the following solution for $p(t)$

$$p(t) = \frac{-e^{2\sqrt{3}T}z_0}{2 + \sqrt{3} + (\sqrt{3} - 2)e^{2\sqrt{3}T}} e^{-\sqrt{3}t}$$

$$+ \frac{z_0}{2 + \sqrt{3} - (2 - \sqrt{3})e^{2\sqrt{3}T}} e^{\sqrt{3}t}$$

▶ and the optimal input is then given by (A.10).

INDEX